Micro- and Opto-Electronic Materials, Structures, and Systems

Series Editor
E. Suhir
University of California, Santa Cruz, CA, USA

More information about this series at http://www.springer.com/series/7493

Jasper Van Hoorick • Heidi Ottevaere
Hugo Thienpont • Peter Dubruel
Sandra Van Vlierberghe
Editors

Polymer and Photonic Materials Towards Biomedical Breakthroughs

Springer

Editors
Jasper Van Hoorick
Polymer Chemistry & Biomaterials Group
Centre of Macromolecular Chemistry
Department of Organic and Macromolecular Chemistry, Ghent University
Ghent, Belgium

Brussels Photonics
Department of Applied Physics and Photonics
Vrije Universiteit Brussel
Brussel, Belgium

Hugo Thienpont
Brussels Photonics
Department of Applied Physics and Photonics
Vrije Universiteit Brussel
Brussel, Belgium

Sandra Van Vlierberghe
Polymer Chemistry & Biomaterials Group
Centre of Macromolecular Chemistry
Department of Organic and Macromolecular Chemistry, Ghent University
Ghent, Belgium

Brussels Photonics
Department of Applied Physics and Photonics
Vrije Universiteit Brussel
Brussel, Belgium

Heidi Ottevaere
Brussels Photonics
Department of Applied Physics and Photonics
Vrije Universiteit Brussel
Brussel, Belgium

Peter Dubruel
Polymer Chemistry & Biomaterials Group
Centre of Macromolecular Chemistry
Department of Organic and Macromolecular Chemistry, Ghent University
Ghent, Belgium

Micro- and Opto-Electronic Materials, Structures, and Systems
ISBN 978-3-319-75800-8 ISBN 978-3-319-75801-5 (eBook)
https://doi.org/10.1007/978-3-319-75801-5

Library of Congress Control Number: 2018936508

Printed on acid-free paper

This Springer imprint is published by the registered company Springer International Publishing AG part of Springer Nature.
The registered company address is: Gewerbestrasse 11, 6330 Cham, Switzerland

Preface

To date, strategies in which photonics, engineering and biomaterials are combined to serve a common purpose have been scarcely reported. Up to now, their cross-fertilization has been mainly limited to the development of photo-responsive polymers for various applications including drug delivery.

One specific example of photo-responsive polymers is are photo-reversible polymers. A photo-reversible polymer can be obtained by introducing functional groups. Which can reversibly dimerize upon irradiation with various wavelengths (see Table 1). The resultant materials combine the benefits of strong covalent interactions with the ease of reversible material manipulation for physically cross-linked materials. In addition, these materials do not require the presence of potentially toxic photo-initiators to result in polymer crosslinking. Common photo-responsive moieties include anthracene, cinnamic acid, thymine, stilbene and coumarin, as they are capable of undergoing photo-reversible dimerization upon applying a suitable UV-induced trigger.

Generally, $[4\pi + 4\pi]$- or $[2\pi + 2\pi]$-cycloadditions occur depending on the applied moiety (see Table 1 and Fig. 1). When applying coumarin, dimerization occurs through the transfer of electrons from one molecule (i.e. the donor) to a second (i.e. the acceptor). This transfer typically occurs from the most energetic electrons in the donor (i.e. the electrons that are present in the highest occupied molecular orbital (HOMO)) to the lowest unoccupied molecular orbital (LUMO) of the acceptor. When light with a wavelength of above 350 nm is applied, an electron can transfer from the HOMO of the alkene in coumarin to the LUMO. However, the development of photo-responsive biopolymers such as gelatin has not received much attention.

Photonics also play a crucial role in polymer processing for photo-responsive polymers. For example, laser-based systems apply photo-polymerization to introduce crosslinks in polymer materials, or to generate polymer chains in a monomer solution. More specifically, stereolithography and the more advanced two-photon polymerization (2PP) method of additive manufacturing (AM) are two well-known examples where polymers and light meet.

Two-photon polymerization is a cutting-edge example of a laser-based technique applying light to initiate the polymerization of a photo-sensitive material, thereby

Table 1 Overview of compounds able to undergo photo-reversible dimerization by application of a suitable UV trigger with (from left to right) stilbene, thymine, coumarin, cinnamic acid and anthracene

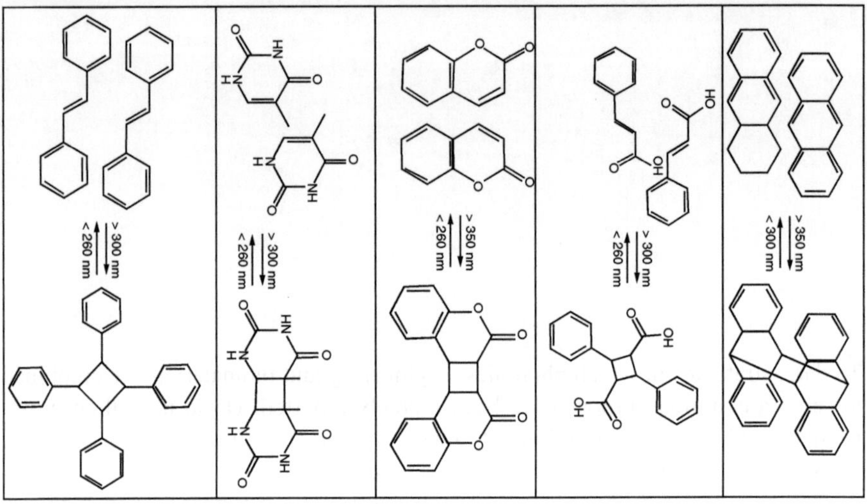

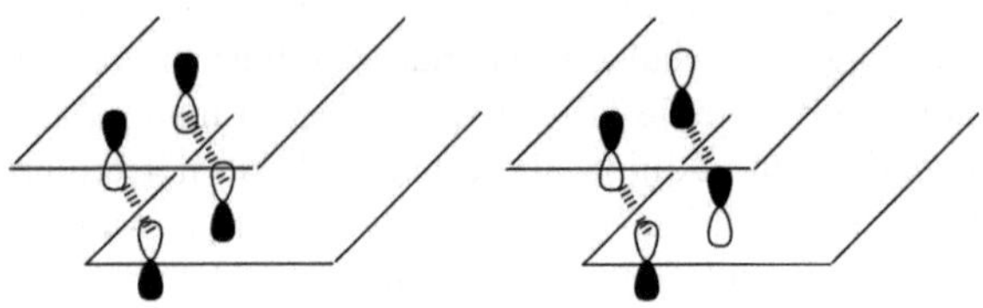

Fig. 1 Orbital symmetry required to enable suprafacial [$2\pi + 2\pi$]-cycloaddition of two alkenes

enabling 3D structuring. Among others, this technique enables the processing of hydrogels in a very controlled manner. Furthermore, as mild processing conditions are applied to enable crosslinking, a broad range of materials including thermally unstable materials can be processed. The latter opens up perspectives to produce biodegradable and cell-interactive scaffolds, which are functionalized with proteins or other biological compounds. Interestingly, the fabrication of materials encapsulating cells has also been performed in order to improve the cell seeding efficiency.

2PP differs from other laser-based AM techniques because it applies the principle of two-photon absorption for the generation of micro- and nanostructures inside a polymerizable solution. This is made possible by the application of a tightly focused, femtosecond pulsed laser beam. These laser pulses have wavelengths in the near-infrared range as opposed to traditionally applied UV exposure in conventional light based AM techniques. When focused correctly, the 2PP setup enhances the probability for an initiator molecule to simultaneously absorb two or more long wavelength photons. This can result in energy comparable to the absorption of one short wavelength photon and thus induce a polymerization reaction confined to the focal spot. As a result, the technique is described using two-photon polymerization or multi-photon polymerization. This is where technology differs from conventional stereolithographic processes. Conventional methods result in polymerization occurring in the entire area where the laser beam penetrates the solution instead of being confined to the focal spot.

When applying 2PP, the photo-initiator concentration should also be taken into account, as these compounds often show some cytotoxicity, which negatively influences cell viability. However, in general, the unreacted photo-initiator can be removed by incubating the generated structures in an appropriate solvent.

Since 2PP does not operate in a layer-by-layer fashion, it benefits from having nearly no geometrical restrictions when producing a structure. The movement of the focal point through the solution can be easily traced and processed through the solidification of the material, resulting in a 3D drawing. 2PP exhibits a higher resolution than other SFF techniques, even beyond the diffraction limit because of this feature. This is achieved by controlling the number and the energy of the applied laser pulses so that they barely overcome the polymerization energy threshold, which results in nanometer precision.

2PP is able to mimic the extracellular matrix (ECM) to a greater extent than other techniques, which increases cell proliferation.

Similar to other SFF technologies, being able to perfectly control the design of the scaffolds enables them to exhibit regions with varied mechanical properties.

Unfortunately, the 2PP technique possesses a disadvantage when compared to alternative SFF techniques. When constructing large scaffolds, it requires significantly more time, while scaffold dimensions still remain limited to date. Additionally, the technology is still relatively young, resulting in few commercially available set ups and concomitant high associated costs.

The above overview is not exhaustive but highlights the versatile applications that result from the combination of light and materials. This book attempts to overview recent biomedical breakthroughs in which polymers, photonics and their combinations were crucial. In addition, the chemical and engineering aspects accompanying these biomedical breakthroughs have been outlined.

Ghent, Belgium Jasper Van Hoorick
Brussel, Belgium Heidi Ottevaere
Brussel, Belgium Hugo Thienpont
Ghent, Belgium Peter Dubruel
Ghent, Belgium Sandra Van Vlierberghe

Contents

Part I
Material Development and Processing

Chapter 1
Development and Characterization of Photoresponsive Polymers

Florica Adriana Jerca, Valentin Victor Jerca, and Izabela-Cristina Stancu

1.1 Introduction

Synthetic polymeric materials capable of responding to external stimuli represent one of the most exciting scientific areas of commercially emerging applications. While there are many challenges facing this field, there is a great deal of opportunities in design, synthesis, and engineering of stimuli-responsive polymeric systems, and Mother Nature is an endless supplier of inspiration [1–3]. This book chapter is focused on reviewing the developments made in the field of *synthetic photoresponsive polymers* that undergo physical changes in response to a light stimulus. The control over the physical properties of a polymer system by light is obviously a highly desirable advantage. *Light* is a particularly interesting stimulus that can be controlled spatially and temporally with great ease and convenience from the exterior; therefore, this topic was countless times reviewed [4–8]. The constant scientific interest toward the highly developed experimental techniques in polymer science, which provides today's insight into polymer's structure and optical properties,

F. A. Jerca
Centre for Organic Chemistry "C. D. Nenitzescu", Romanian Academy, Bucharest, Romania

V. V. Jerca (✉)
Supramolecular Chemistry Group, Department of Organic and Macromolecular Chemistry, Ghent University, Ghent, Belgium

Centre for Organic Chemistry "C. D. Nenitzescu", Romanian Academy, Bucharest, Romania
e-mail: valentinvictor.jerca@ugent.be

I.-C. Stancu (✉)
Faculty of Medical Engineering, University Politehnica of Bucharest, Bucharest, Romania

Advanced Polymer Materials Group, University Politehnica of Bucharest, Bucharest, Romania
e-mail: izabela.stancu@upb.ro

J. Van Hoorick et al. (eds.), *Polymer and Photonic Materials Towards Biomedical Breakthroughs*, Micro- and Opto-Electronic Materials, Structures, and Systems, https://doi.org/10.1007/978-3-319-75801-5_1

helped to make the photoresponsive materials invaluable assets in everyday life. The field of photoresponsive polymers is a vast domain of research that covers development of polymeric materials for high-tech industry, such as photonics [9–11], biotechnology [12–14], or telecommunications [15]. One obvious example of a practical application of photoresponsive materials relies in the use of sunglasses that darken on exposure to bright sunlight and regain their transparency in darkness or low light intensity. In green high-tech applications, photoresponsive polymers are being used to harvest solar energy and store it for significant amounts of time. Other commercially available products of such polymer systems can be found as toys, cosmetics, and clothing.

Considerable widespread research is dedicated to photoresponsive polymers in the form of original research articles, reviews, books, and book chapters, ranging from fundamental studies to emerging applications. There have been reports on photoresponsive polymer systems that convert light into mechanical energy that enables the possibility of bending, flexing, swelling, contraction, and motion [16–20]. However, to start with early developments in this area, the most debated featuring applications of photoresponsive polymers are of course related to the nonlinear optical (NLO) phenomena [11, 21]. A number of several important discoveries made in the 1990s have had a particular impact on the NLO field. These include the surface relief grating (SRG) that can be inscribed on photochromic-based polymers using an interference pattern as a result of photoinduced mass transport [22, 23] and the photochemical phase transition from liquid crystalline (LC) to isotropic (order–disorder) phase because of the perturbation effect arising from the photoisomerization process [24–26]. Most of the research efforts in the 1990s mainly dealt with the physical and optical properties of chromophore-containing polymers and liquid crystals which brought important contribution for the optical signal processing [27], all optical switching to nonlinear optical imaging [28], to reversible optical storage [29, 30].

Today, the research field of photoresponsive polymeric materials remains still extremely active, thanks to important new directions and developments over the last 10 years in the biomedical field, which witnessed exciting progress in cellular biology [31–33], tissue engineering [34, 35], and drug delivery [4, 35–38]. The abundance of research articles in the photoresponsive polymers domain is driven by its enormous potential to contribute to improving health and life quality which are among the most important human priorities in our world. For this reason, it is necessary to reveal the actual challenges that fellow researches need to overcome with the design of the materials, their formulation, and properties and to discuss future research directions. A comprehensive illustration of the existing topographic categories is given in Fig. 1.1, which includes most important and most investigated photoresponsive polymer systems in both biomedical and NLO field.

Although in nature we can find several inspiring examples of how to approach and design functional photoresponsive materials, creating such controllable synthetic systems that respond to a light stimulus in a predictable fashion still represents a great challenge. First of all, it is important that the photoresponsive polymeric-based biomaterials to overcome the restricted mobility within the

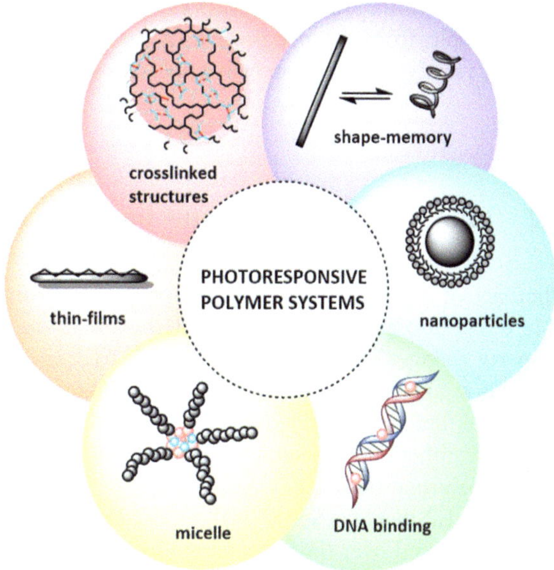

Fig. 1.1 Schematic representation of the dimensional existing variety of structured photoresponsive polymer materials, which range from shape-memory materials [39] to NLO materials (in the form of assemblies in thin films or multilayered films made of different polymers [40]); to crosslinked structures [41], micelles, and particles and their assemblies in solutions for drug delivery [42]; and to DNA/RNA binding for bioactivity [43]

network, while maintaining their mechanical integrity, without imposing limits on obtaining the photoresponsiveness. Another argument in this regard lies in mimicking biological systems where structural and compositional gradients at various length scales are necessary for orchestrated and orderly responsive behaviors. Significantly greater challenges exist when designing polymeric materials to exhibit biocompatibility and biodegradability, whereas the degradation products should be nontoxic. In addition to all these arguments, it is very important to have a highly reproducible composition via synthesis, functionalization or crosslinking when preparing any type of synthetic material.

To address all these problems regarding the structural nature, several photoresponsive systems have been developed over time (see Fig. 1.1), with the majority of studies dealing with hydrogel materials [35, 44–46], polymeric solutions, micelles and core–shell particles [4, 47–52], and to some extent polymeric solids [53]. To bring into consideration one of the most significant matter related to the photoresponsiveness of these polymer systems is when using UV irradiation as a stimulus. This stimulus is regarded as a relatively straightforward and noninvasive process to induce light-responsive behavior. However, most biological systems can suffer damage upon this type of irradiation. This is a strong argument which can limit the biomedical applicability of several photoresponsive polymeric systems. Therefore, the constant need to develop new photoactive molecules and improve the existing photoresponsive polymeric systems is imperative. At the "heart" of functional photoresponsive systems underlies the photoactive molecules. They can be incorporated into polymers in various topological configurations, such as side groups of linear polymers, within the backbone of linear macromolecules or as crosslinkers within branched networks. In this book chapter, we aim to approach the present topic from the perspective of the photoactive species that make up the existing variety of photoresponsive polymer systems and to highlight their function in use.

1.2 Photoresponsive Systems based on Photoactive Groups

Photoactive molecules play a pivotal role within photoresponsive systems, being able to capture an optical signal and convert it via a photoreaction, into a useful property change. This chapter's purpose is to cover the wide variety of photoactive molecules and to highlight the photochemical transformations they undergo. Therefore, here are included numerous classes of molecules that respond to light, which are either photochromic molecules or just photoactive, that have been used to trigger the properties of polymeric systems.

Figure 1.2 displays some of the most studied photoactive molecules that have been engineered over time to respond to light, ranging from ultrashort wavelength lasers to near-infrared light, and have been embedded into functional polymeric systems. Before going into detail about the new developments in the biomedical field, a classification of most important photoactive molecules is necessary. The polymer systems incorporating these photoactive molecules become functional and useful depending on the working principle they obey in response to light.

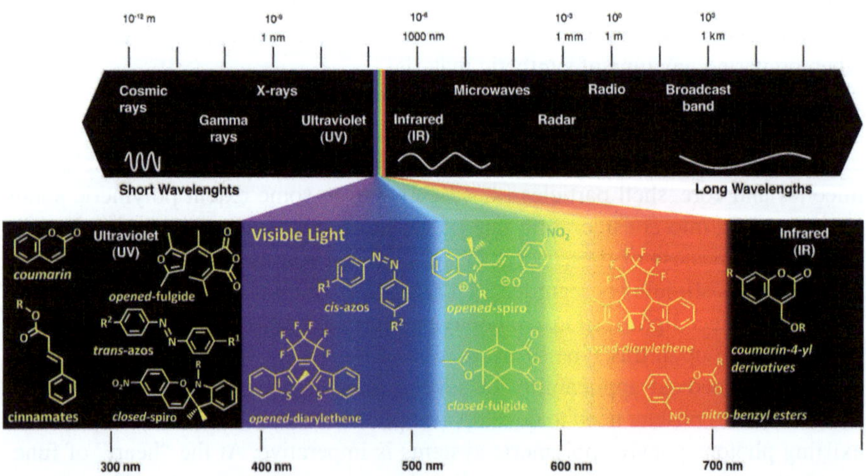

Fig. 1.2 Large-scale view over the variety of photoactive molecules that show typical absorption wavelengths in the broad absorption band that are frequently employed to photoregulate polymeric systems

1.2.1 Origin, Definition of Photochromism, and Basic Operation Principle

The phenomenon of photochromism was first observed by Fritzsche in 1880, who noted a reversible color change in a solution of tetracene upon sunlight exposure, but it took more than 70 years for the scientific community to define this phenomenon and call it photochromism [54]. Photochromic reactions are reversible and unimolecular processes that involve the transformation upon irradiation with light between a thermodynamically stable configuration of a species A to corresponding species B [55]. The species B can return to the ground state through thermal or photochemical processes. If the photogenerated isomers are unstable and revert thermally to their ground state in the dark, then they are termed T-type (thermally reversible type, e.g., spiropyrans and azobenzenes). The photogenerated isomers that are thermally irreversible but photochemically reversible are termed P-type (e.g., fulgides and diarylethenes). In addition to a color change in some cases, the intramolecular ring opening/closing and *cis–trans*-photoisomerization of the two isomeric forms induced by the absorption of light exhibit also a substantial change in the absorption spectra.

The photoresponsive systems typically are composed of a polymeric network and a photoactive moiety, usually a photochromic chromophore as the functional part. The optical signal is first captured by the photoactive molecules. Then, it is converted to a chemical signal through a photoreaction such as isomerization, cleavage, or dimerization, and this processed signal is transferred to the functional part of polymers to tamper with its properties.

1.2.2 Type of Photoactive Molecules and Photoreactions

The numerous photoactive molecules used as a point of origin for polymeric systems include, but are not limited to, azobenzenes [7, 10, 56–60], spiropyrans [61–63], spirooxazines [64–66], fulgides [67], coumarins [68–73], and *o*-nitrobenzylesters [74–79], as outlined in Fig. 1.3. Although all aforementioned classes are light-responsive molecules, not all of them are photochromic molecules. To cover the large availability of photoactive molecules, regardless of their molecular structure, they will be discusses based on the mechanism they obey under irradiation with light. Thus, the photoactive molecules fall in three categories that include photoisomerization reactions, photodimerization reactions, and photocleavage. These dynamic photoreactions have been exploited to generate considerable changes in the optical, chemical, electrical, and bulk properties in the systems that incorporate them.

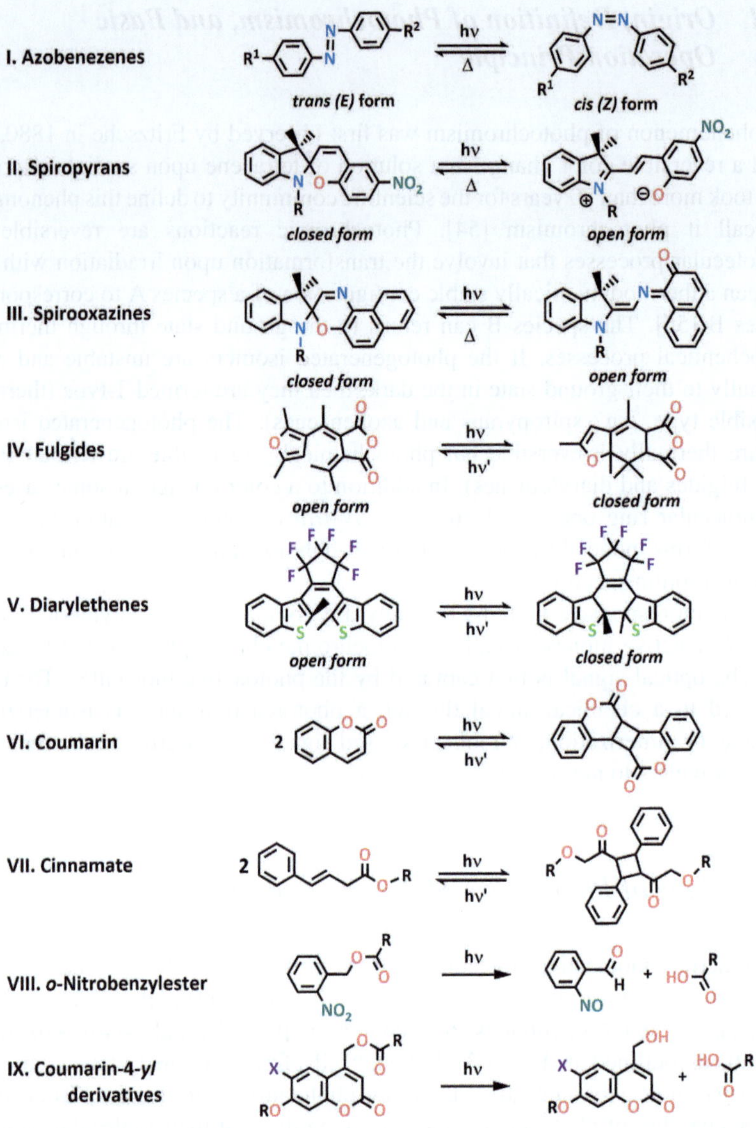

Fig. 1.3 Most widespread photoactive molecules used to control polymer's properties

1.2.2.1 Photoisomerization

The photoisomerization processes are reversible and repeatable and are regarded as the most "clean" photoreactions in chemistry, since only two absorbing species are formed during photoisomerization. This property makes the photochromic molecules very attractive, and thus, photoresponsive polymeric materials are of significant scientific, technological, and commercial interest because photoconversion and photoreversion modulate a multitude of physical properties not just color but also geometrical shape, dipole moment, refractive index, birefringence, conductivity, magnetism, hydrophilicity, hydrophobicity, adhesion, and so forth [1, 7, 8, 58].

The *azobenzene* is perhaps the most investigated photochromic molecule that can undergo a reversible photoisomerization reaction from a stable *trans* to a metastable *cis* conformation. As shown in Fig. 1.3 I, the reversible *trans–cis* isomerization of azobenzenes can be described as a geometrical isomerization. Most of the times their interconversions are not visualized as a distinct color change, as in the case of other photochromic transformations; however, the difference between the isomers is visible by their different λ_{max} values. Due to the numerous reports on this topic, we dedicate an entire chapter to expound the features and particularities of these molecules. The versatility of these photoactive compounds in synthesis and design to address the most recent challenges in the biomedical field will be addressed.

Spiropyran is another well-known photochromic switch. The spiropyran–merocyanine transformation relies on UV-induced (approximately 360 nm) photolysis of the Cspiro–O bond in spiropyran (colorless closed form) to generate the intensely colored open form, merocyanine. Taking advantage of this photochemical transformation, the spiropyran and spirooxazine molecules (Fig. 1.3II, III) were widely used to control the nonlinear optical properties [66], to structure and function biomaterials with light, and to obtain photoresponsive hydrogels and micelles [33, 80, 81]. Similarly, the fulgides and diarylethene derivatives, which are thermally stable (Fig. 1.3IV, V) but photochemically reversible, have also been used for the functionalization of polymers. Both classes have been used in optical memory, photooptical switching, and displays and have been extensively reviewed elsewhere [82, 83].

1.2.2.2 Photodimerization

Light-induced reversible dimerization is another strategy to confer photoresponsive properties to polymeric systems. Dimerization describes the process in which two previously unbound molecules are covalently coupled to each other. Several compounds have been reported having such reversible dimerization properties upon exposure to light including cinnamylidene acetate [84], nitrocinnamate [85], and anthracene [86]. However, cinnamic acid and coumarin derivatives are the most frequently employed molecules that can undergo reversible photodimerization (Fig. 1.3VI, VII). Both have been incorporated as functional components into many types of photoresponsive polymeric systems, such as light-induced reversible crosslinkers for in situ modification of hydrogels [71] or transient stabilization of micelles [87] or used as crosslinking points in photoinduced self-healing materials [88, 89]. Shape memory polymers that can switch between temporary and permanent geometries upon exposure to light were synthesized by incorporation of cinnamate or coumarin groups [90]. Light-induced recovery of permanent geometries in shape-memory polymers using reversible network formation/cleavage via cinnamate-based cross-links represents an orthogonal actuation cue compared to formally temperature changes [17]. Although photodimerization can crosslink and cleave polymer chains using light, some practical biomedical applications remain elusive, due to the UV absorbing reversible process (260 nm) of these structures, which can cause cell damage. Cinnamate- and coumarin-bearing polymers could have more promise when used as photoresists for microelectronic fabrication or as environmentally benign materials that can be decomposed upon light exposure after the expected life cycle of the material.

1.2.2.3 Photocleavage

Photocleavage of photoactive protecting groups is another interesting approach to induce photoresponsiveness in polymers. The concept of incorporation photolabile groups into polymer systems originates from a study that aimed to engineer selective biological activity in signaling molecules [91]. To restore the native biological function of the small molecule, light-induced cleavage of photoactive groups was used [92]. The *o*-nitrobenzyl group is one of the most useful photolabile compound for photoresponsive polymeric systems used for biomedical purposes (Fig. 1.3VIII). The biocompatibility of the *o*-nitrobenzyl moiety has been demonstrated for the natural endothelium [93]. However, some studies show that upon photoirradiation, the cleavage of the *o*-nitrobenzyl group yields a cytotoxic nitrosobenzaldehyde derivative, which significantly inhibits the proliferation of cells under standard in vitro conditions [94]. Despite of this drawback, this molecule affords by substitution to adjust polymer's photochemical properties, such as red shifting the wavelengths for photolysis, quantum yield, absorbance maximum, and extinction coefficient [95]. Thus, this photocleavage reaction has been used extensively to fragmentize side chain, main chain, or end groups in numerous polymer architectures in an ordered fashion and to control supramolecular interactions by changing the chemical properties of interacting molecules [75, 96–98]. Another interesting characteristic of the *o*-nitrobenzyl photoactive derivatives is that they can be tailored to undergo cleavage through a nonlinear optical process that requires simultaneous absorption of two photons [99]. Since the two-photon process can be achieved with near-infrared (NIR) radiation which is less absorbed by the living tissues than UV radiations, it is beneficial for in vivo applications. Two-photon absorption provides a very promising path for preparing photoresponsive polymers with three-dimensional microstructures used in regenerative medicine and controlled release [47, 79].

Coumarin-4-ylmethyl (Fig. 1.3IX) and its derivatives can also serve as protective groups for selective photocleavage and have been showed to exhibit larger two-photon absorption cross section compared to *o*-nitrobenzyl derivatives [34, 100]. Alcohols [101], phenols [101], and amino acids [102] have been conjugated with coumarin-4-ylmethyl through photolabile bonds to give photoresponsiveness to the polymeric system incorporating them. Exhibiting a higher penetration depth compared to the *o*-nitrobenzyl derivatives, the risk to damage the cells and tissues is diminished. Also, coumarin derivatives are well known for their biocompatibility and natural biodegradability, therefore, they are more suitable candidates for the biomedical field [73].

Photolysis of *p*-hydroxyphenacyl groups [103], acetal–ketal protecting groups [104], and triphenylmethane derivatives [105] have also been used in the related fields to confer photoresponsiveness to polymeric systems.

1.3 Figures of Merit for Photoresponsive Azobenzenes

Azobenzene, with its two phenyl rings separated by the azo bond, is the point of origin for a broad class of aromatic azo-derivatives. The most interesting property of these azo compounds is the induced and reversible isomerization of the azo bond between the thermally stable *trans* configuration and the *meta*stable *cis* form. Most azobenzenes can be optically isomerized from *trans* to *cis* with light, anywhere within the broad absorption band. Once formed, *cis* isomers will thermally reconvert back to the stable *trans* state within a timescale dictated by the substitution pattern, which depends greatly on its interaction with the surrounding medium. This "clean" photochemistry is the most important feature of azobenzene and offers a reversible control over a variety of chemical, electronic, and optical properties [7, 27, 106]. This light-induced interconversion is accompanied by a large geometrical transformation from the extended *trans* configuration to the three-dimensional and more compact *cis* isomer, which allows to alter the natural structural order of systems incorporating them to substantially tune up the strength of the host–guest interactions [57, 107].

Depending on the absorption wavelength, the azobenzene chromophores were divided into three general classes, as described by H. Rau early in the literature [108]. The azobenzene- type molecules which absorb in the UV (330–370 nm) range and their *cis* configuration can be stable for days in the dark, the amino-azobenzenes that have an intermediate lifetime and a slight red shift of the *trans* absorption band, and the pseudo-stilbenes which exhibit a very fast thermal reconversion and a far-red-shifted maximum of absorption. These features gives the azobenzene class a significant advantage over the previously enumerated photoactive compounds in Fig. 1.3, as the chemical substitution pattern has a large effect on the photophysical properties [58, 59, 109, 110]. In addition, azobenzenes can be photoisomerized on a timescale of microseconds down to sub-nanoseconds, reversibly 10^5–10^6 times without exhibiting side reactions.

The recent contributions in the photochemistry of azobenzenes show just how versatile it is the nature of this class in design and synthesis. In Fig. 1.4 are illustrated two classical examples of azo-molecules, the parent azobenzene and Disperse Red 1 (DR 1), by comparison with the latest designs. While DR 1 brought large contribution in the photochemistry of azobenzene, exhibiting exceptionally high nonlinear optical activity [107, 111], the other azo-derivatives were synthesized to

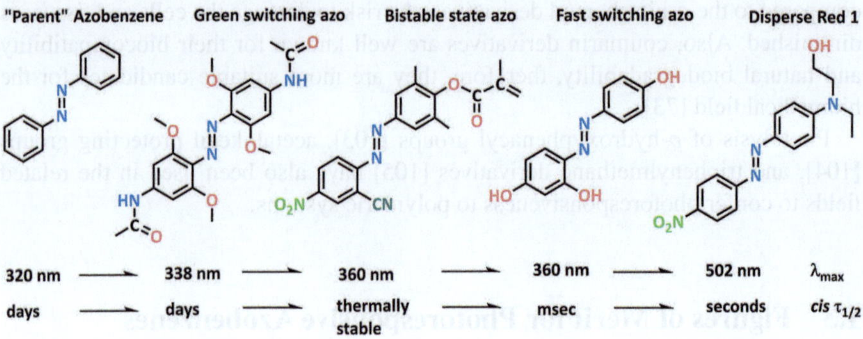

Fig. 1.4 Examples of azobenzene derivatives tailored over time to meet specific photophysical requirements. Ranging from the UV absorbing "parent" azobenzene to an aminoazobenzene with two electron-donating groups that isomerize to *cis* with visible light for biomedical purposes, to a thermally two-state azobenzene derivative for optical storage data, to fast photoswitching azo-derivative for real-time information-transmitting data, and to a strong electron push–pull pseudo-stilbene DR1, which possess a strong red-shifting absorption that is visibly useful in nonlinear optical imaging

face the present requirements in both NLO and biomedical field [58]. Based on rational design, azo-derivatives that *trans–cis* isomerize under green light irradiation have been reported by Woolley's group [60]. New azo-derivatives were designed to exhibit fast relaxation times, based on solvent interactions for molecular photoswitching by Velasco's group [112]. Elegant examples of thermally stable *cis* configurations in azo-derivatives were reported by Jerca's group [56, 113]. The incorporation strategy is the key to exploiting azobenzene unique behavior, and one of the most attractive methods for incorporating azobenzene into functional materials is covalent attachment to polymers. The resulting materials will benefit from the inherent stability, rigidity, and processability of polymers, in addition to the target photoresponsive behavior of the azo moieties. However, there have been increasing efforts toward the development of new diverse functional materials and azo-derivatives, and the last 10–15 years witness emerging applications other than information storage and switching. In the next subsections, a brief description of azo-derivatives contribution to photopharmacology and to biological systems will be made.

1.3.1 Azobenzenes in Photopharmacology

One of the most important area in the biomedical field that witnessed important progress is the *photopharmacology*, which has been formulated by Feringa group as a viable alternative to optogenetics [114]. Photopharmacology is a complementary branch of research of photoresponsive drugs whose activity can be regulated with light in the so-called near-infrared phototherapeutic window (650–900 nm).

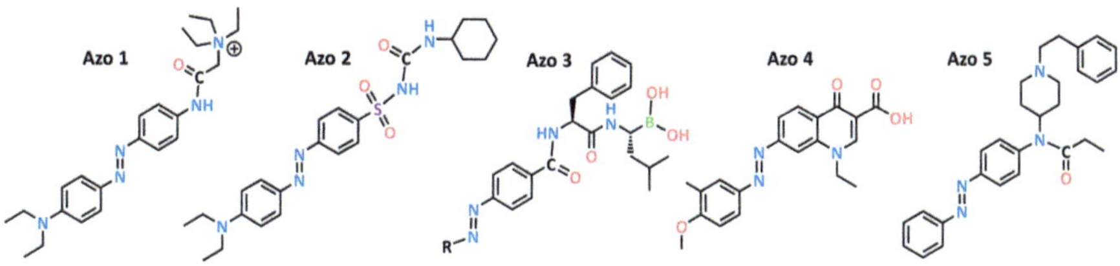

Fig. 1.5 Photoresponsive bio-inspired azo-derivatives with potential application in the medical field: (1) ophthalmology, (2) diabetes, (3) cancer therapy, (4) antibiotics, and (5) analgesics

However, Erlanger and coworkers pioneered the concept of *photopharmacology* around 1970, after reporting the control of acetylcholinesterase (AChE) activity with light by using azobenzene-based inhibitors [115]. Currently, photopharmacology is at the stage of defining and evaluating the molecular targets, solving the problems arising from design and synthesis, supported by the results of in vitro studies and general cellular toxicity. Many research groups have recently formulated bio-inspired molecular photoswitches, based on the azobenzene scaffold, which can be operated in or near the therapeutic window (see Fig. 1.5). A selection of studies has been chosen to highlight the advances made in the field of ophthalmology, the diabetes research, and cancer therapy. However, they will not be discussed in detail, because they make the topic of another review [114].

The Trauner group aimed at fighting *retinitis pigmentosa* and age-related macular degeneration, which are both diseases that cause blindness through the death of the rod and cone photoreceptors [116–121]. Through sustained efforts, Trauner and coworkers managed to formulate a fast-switching azobenzene with a red-shifting absorption wavelength of 450–550 nm (see Fig. 1.5 Azo 1), by introduction of a strong electron-donating diethylamine group to the system [117]. This fast photoswitching process enabled a selective interaction with regions to which the cell death was constrained and also lead to a profound effect observed on the photoreceptor-degenerated retinas in wild-type or triple knockout (TKO) mice [116].

The need to control insulin levels in a precise temporal manner makes diabetes an excellent target for photopharmacology. Recently, Trauner and coworkers published several studies that showed that both insulin release and pancreatic beta cell function could be controlled with UV light [122, 123], using azo-derivatives as photoswitches (see Fig. 1.5 Azo 2). However, limitations of the presented designs include the use of short-wavelength light (400 nm) and the lack of thermal stability of *cis* configuration, together with a small difference in the activity between the *trans* and *cis* isomers. A bistable azo with longer wavelength light, combined with larger differences in potency between the isomers, would drive this elegant proof-of-concept closer toward clinical application.

The first reported examples of azo-based photoswitchable anticancer drugs have been formulated and reported by Feringa group [124]. Several light-responsive inspired derivatives were synthesized (see Fig. 1.5 Azo 3) based on a chemotherapeutic proteasome inhibitor structure, which has been proven successful against multiple myeloma and mantle cell lymphoma [125]. The photoswitchable compounds were found to show two- to threefold differences in activity, with different selectivity toward the different active sites of the proteasome. The different inhibitors also showed activity in MTT cytotoxicity assays on HeLa cells. However, the use of UV light, especially for delivery to the bone marrow, is a drawback toward clinical applications. Moreover, a larger difference in activity between the two isomers would be beneficial in this case also. Other examples of azo-derivatives used in anticancer research have been reported by other independent groups [126–128]. The studies included the development of combretastatin A4 analogues, which represent powerful examples of photoresponsive drugs that can be comprised in the photopharmacology branch. Last, but not least, remarkable examples of reversible control over antibiotics (see Fig. 1.5 Azo 4) [129], anesthetics, and analgesics (see Fig. 1.5 Azo 5) [130, 131] with light have been reported in recent years, offering an exciting opportunity for developing many interesting possible applications.

1.3.2 Azobenzene Functionalized Biomacromolecules

Biomolecules have been an exciting area of research in the application of photoresponsive materials [58]. Biomacromolecules functionalized with azobenzene such as deoxyribonucleic acid (DNA) and peptides can undergo substantial conformational changes upon isomerization, leading to designed and reproducible changes in biomaterial function [132–135]. These biomolecules will undergo predictable folding or unfolding in a biological context, if a trigger such as light is used for disruption. The Makoto Komiyama's group reported for the first time the concept of chemical incorporation of an azobenzene derivative that could modulate the melting temperature of DNA throughout the double helix disruption [136]. After the acceptance of these results by the large scientific community, other reports were further expanded toward photoregulation of hybridization and transcription [134], as schematically shown in Fig. 1.6.

Potential applications of these systems in drug delivery were investigated by W. Tan's group who have exploited the idea of disrupting a DNA duplex with light to release small drug molecules, proteins, and nanoparticles or other cargos [137, 138]. In an elegant study, they described the development of a photoresponsive DNA-crosslinked hydrogel, whereas photosensitive azobenzene moieties were incorporated into DNA strands as crosslinkers, which can be photoregulated by two wavelengths with a reversible sol–gel conversion. This photoinduced conversion can be further utilized for precisely controllable encapsulation and release of multiple loads. To prove the concept, different materials, fluorescein, horseradish peroxidase, and gold nanoparticles, were encapsulated inside the gel while irradiating at 450 nm and then released under irradiation at 350 nm. Further in vitro experiments were carried out to deliver the chemotherapy drug doxorubicin in a similar manner. The results show a net release rate of 65% within 10 min, and the

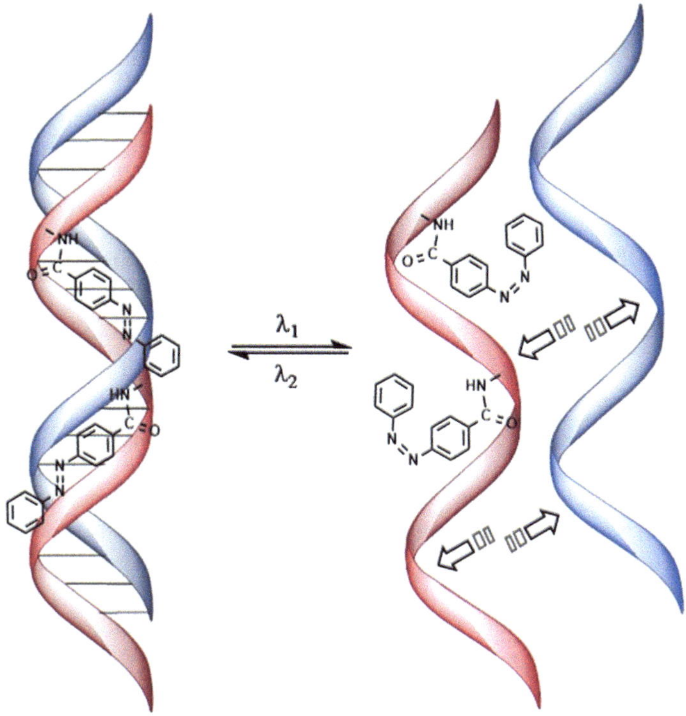

Fig. 1.6 The photoregulation of DNA hybridization as envisioned by Komiyama et al., starting from a phosphoramidite monomer bearing an azobenzene synthesized from D-threoninol. Using this monomer, azobenzene moieties can be introduced into oligodeoxyribonucleotide (DNA) at any position on a conventional DNA synthesizer. When the azobenzene takes a *trans*-form, a stable duplex is formed. The duplex can be dissociated into two strands after isomerization to its *cis*-form by UV light irradiation (360 nm). The duplex is reformed on photoinduced *cis–trans* isomerization (>400 nm)

released drug maintained its therapeutic effect. This hydrogel system provides a promising platform for drug delivery in targeted therapy [137].

Another interesting example of drug delivery system was designed to incorporate a photoisomerizable azo group into the tails of a phosphocholine-based lipid [139]. These liposomes were formulated to accommodate the active loading of drugs (acridine orange and doxorubicin), allowing for high drug encapsulation. When in the *trans* conformation, the lipid bilayer was well packed and relatively stable with little release observed. However, in response to visible light in the region of 470 nm, isomerization from *trans* to the *cis* conformation of the azo group was induced, and the disruption of the packed membrane accompanied with release of the two encapsulated model drugs was observed.

Despite the extensive study of azobenzenes, this class of molecules still has great challenges to overcome. Starting from how to design longer wavelength switches that exhibit large conformational changes and excellent switching behavior to translation of these changes into large and productive changes in biomolecular structure and function.

In the next sections, the most important contributions that azobenzene polymeric systems and the other photoactive polymers have brought in the drug delivery field and bioimaging will be expounded.

1.4 Photoregulated Drug Delivery of Photoresponsive Polymeric Systems

In the past decades, polymeric micelles have emerged as one of the most promising technology platform for targeted and controlled drug delivery. Micelles play a critical role in biology, offering a dynamic barrier separating two different environments inside living organisms. To this end, various "smart" polymeric nanocarriers that release drugs in response to an internal or external stimulus such as pH, redox, temperature, magnetic, or light have been actively reviewed [4, 5, 37, 42, 140]. These stimuli-responsive micelles have demonstrated gradually improved in vitro and/or in vivo drug release profiles [6]. To further improve drug release performances, novel dual and multi-stimuli-responsive polymeric nanocarriers that respond to a combination of two or more signals have been developed. However, they make the topic of another review [37]. Visible light is a stimulus which provides the least disruptive behavior to biological systems and a high spatiotemporal selectivity with strong dosage control. Because of this property, light-responsive micelles have been investigated for a long time as a viable approach for drug delivery [4].

Hydrogel-based polymeric systems represent another interesting approach to address the drug delivery field [141]. In this type of system, the drug molecules are physically entrapped in the network and are released from the hydrogel matrix by diffusion or by erosion of the mesh of the network [34, 35]. Photoresponsive hydrogels can be easily modified by synthesis to tune their characteristics and can lead to targeted delivery, extension of circulation time, and reduction of toxicity and side effects [35]. Furthermore, hydrogels can be excellent carriers not only for stable low-molecular-weight drugs but also for fragile bioactive macromolecules including proteins, peptides, or mammalian cells [46, 142, 143], endorsing them with a distinct advantage over the micelles.

Both strategies were extensively studied over the years as viable delivery systems, each one with its advantages that arise from synthesis and the use of photoactive molecules. The burst of synthetic polymers that address the biomedical field is due to the development and advances in controlled radical polymerizations (CRPs) that have contributed to the engineering of advanced multifunctional polymeric structures with precise control over architecture, shape, size, surface charge, and functionalization [144–146]. Due to the controlled polymerization techniques, now the development of biodegradable and biocompatible polymers used in medical field is possible [13].

The purpose of this book chapter is to summarize, as much as possible, the approaches described in the literature so far to design functional photoresponsive polymeric systems that can be controlled effectively with light to release drugs or bioactive molecules.

1.4.1 Photoresponsive Micelles

The design and synthesis of this class of photoresponsive systems have been motivated by potential applications for programmable on-demand release for poorly water-soluble chemicals [47]. In the envisioned scenario, the micelle cores are loaded with the hydrophobic payload and self-assemble in aqueous environments due to hydrophobic interactions. Then, the controlled release of the encapsulated agents is being targeted with light. For example, tumor chemotherapy may be more effective if drug molecules are released only after it has been internalized within the cancer cells [147].

Amphiphilic block copolymers are well known to undergo phase separation due to their distinct chemical nature resulting in selective solubility of one block. The self-assembly of amphiphilic polymers into micelles can be explained by the interaction of the hydrophobic and hydrophilic regions with each other as well as with the medium. The hydrophilic shell stabilizes the hydrophobic core by acting as an interface between the aqueous phase and the hydrophobic region. Although there are many reports on amphiphilic block copolymers that can self-assemble in micelles, in this section a selection of studies that include the most significant light-responsive block copolymers investigated for drug release assisted by light will be expounded. Also, a clear-cut separation of the chemical approaches employed for designing micelles responsive to light is needed. Based on the mechanism they obey under light irradiation to photoregulate the structural changes, the light-responsive micelles can be sort roughly into three categories, just like the photoactive compounds they incorporate as schematically illustrated in Fig. 1.7.

1.4.1.1 Shifting Hydrophilic–Hydrophobic Balance

The first approach based on optically shifting the hydrophilic–hydrophobic balance of block copolymers is the most studied one (Fig. 1.7a). Several photochromic molecules have been used in this design, including azobenzenes [148–153], spiropyrans [49, 154–157], diarylethene [158, 159], and diazonaphthoquinone (DNQ) [160, 161]. Most of them display a reversible photoisomerization reaction upon ultraviolet (UV) and visible light absorption, including the *trans–cis* isomerization with azobenzene, the isomerization of spiropyran to merocyanine, and the conversion between ring-open and ring-closed forms of diarylethene.

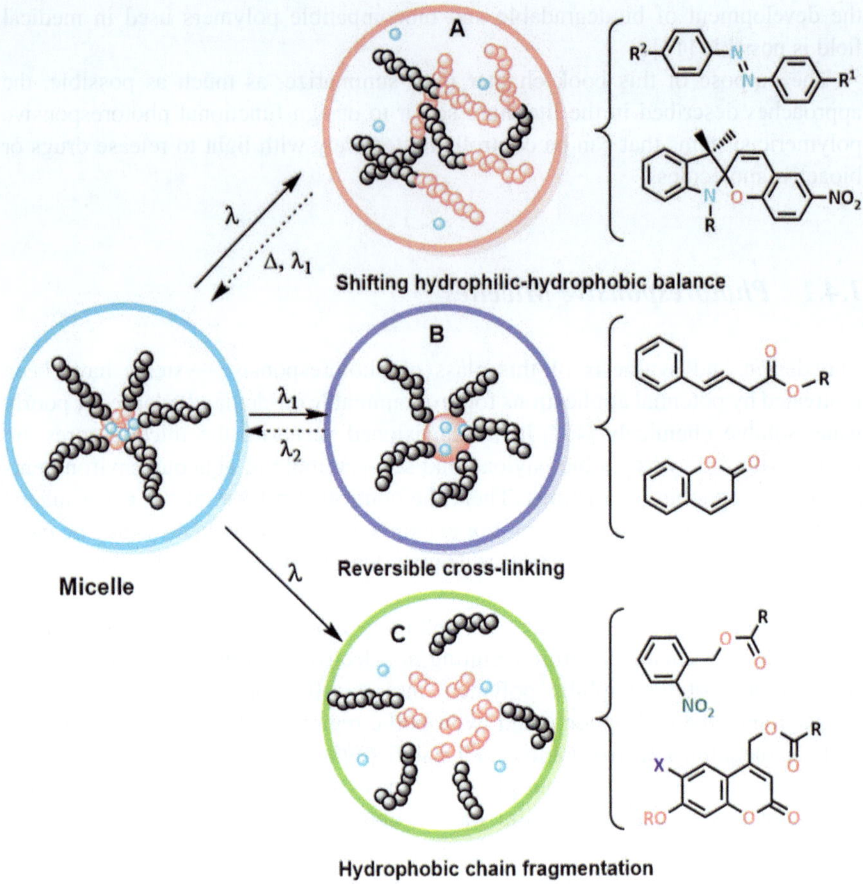

Fig. 1.7 Schematic illustrations for drug release approaches using (**a**) micellar disassembly due to the increase of the hydrophilic character induced by photoisomerization of a small number of photochromic groups, (**b**) reversible or irreversible photoinduced crosslinking in a micelle for core stabilization, and (**c**) photoinduced micelle disruption through main-chain or side-chain fragmentation of the hydrophobic block triggered by removal of photolabile protecting groups

Generally, the photochemical reaction is made to occur inside the micelle, resulting in an increase in the polarity (or water solubility) of the hydrophobic block. This change may shift the hydrophilic–hydrophobic balance toward the destabilization of the micelle and thus lead to its disassembly in aqueous solution accompanied with its cargo release. If the photoreaction is reversible, the initial balance can be restored upon exposure to light at a different wavelength, and the micelle can be reassembled in solution.

Zhao and coworkers exploited this idea and designed micellar systems which can disintegrate upon UV irradiation and reform itself when irradiated with visible light [148, 153]. They synthesized by atom transfer radical polymerization (ATRP) a block copolymer containing one random poly(t-butyl acrylate-co-acrylic acid) sequence connected to a poly-(methacrylate) bearing azobenzene chromophores as side chains. Core–shell micelles could be obtained from this polymer in a dioxane/water mixture depending on the water percentage added to facilitate their formations. The morphology was disrupted upon UV irradiation and reformed after exposure to visible light. However, a more substantial study of shifting the hydrophilic–hydrophobic balance of micelles to control the drug release was reported by Yu's group [150]. In their study, a novel amphiphilic diblock copolymer composed of a hydrophilic poly(ethylene oxide) block and a hydrophobic block copolymerized by azobenzene-containing methacrylate and N-isopropylacrylamide was synthesized using ATRP. The polymeric micelles showed dual responsiveness to heat and light. The temperature change can induce the release of the encapsulated substance; however, the light irradiation can tune the hydrophobicity of the micellar cores with neither disruption of the micelles nor leakage of the encapsulated substance, indicating these micelles are promising in multi-applications.

To enhance the efficiency of reversible photoinduced micellar dissociation–association process, the use of photosensitive spiropyran moieties has been demonstrated [49, 156, 157]. PEO-block-poly(methacrylate), whose methacrylate block bears spiropyran (SP) side chains, were synthesized by ATRP [156]. The photoisomerization is triggered under UV irradiation, and the reverse process is triggered by visible light (620 nm). This process is accompanied by a deep change in the hydrophobic to hydrophilic balance in the core, which causes micellar disruption. Moreover, this system can be successfully used for the encapsulation and release of molecules of interest.

Micelles are known to be dynamic structures which disintegrate themselves below their critical micellar concentration (CMC). In addition, they are sensitive to the nature of the solvent, pH, ionic strength, light, or temperature for the stimuli-responsive designed ones. One way to conserve those micelles in a more robust assembly is to stabilize them by crosslinking.

1.4.1.2 Reversible and Irreversible Crosslinking

Light-induced crosslinking is a convenient route to stabilize micelles, having the advantage to avoid the use of chemical reagents and their unwanted by-products. Early approaches were reported on irreversible micellar crosslinking of the cinnamic esters that undergo [2+2] photo-cycloaddition under UV irradiation [162–164]. An illustrative example of their use in drug delivery systems was formulated by Yotaro's group [165]. In this study, PEO-block-poly(2-(diethylamino) ethylmethacrylate-co-2-cinnamoyloxyethyl-acrylate) were prepared by reversible addition–fragmentation chain transfer polymerization (RAFT). Being pH-responsive block copolymers, due to the amino moiety which undergoes nitrogen quaternization in acidic media, the structures exhibited a water micellar self-assembly stability at a pH higher than 7. However, when the cores of the micelles are crosslinked upon UV illumination due to the presence of the cynamoyl moieties, they maintain their integrity upon pH stimulation. Although the nitrogens from the amino sequence remain quaternizable and sensitive to pH, the acidity cannot disassemble the micelles anymore. Based on this feature, the study showed that 1-pyrenemethanol (hydrophobic dye) can be retained in the hydrophobic core at pH = 10, while in the acidic condition (pH = 3) the polar hydrated core released the dye. Although crosslinking enables a much greater stability of the micelle, the release of encapsulated molecules might not be always easy in crosslinked micelles.

Therefore, the possibility to crosslink micelles in a reversible way is a more feasible approach (Fig. 1.7b) and has been used to photoregulate the release of the molecules in a controlled fashion. In addition, if the crosslinking is reversible, it allows a good robustness of the micelles in the crosslinked state without interfering with the release in the non-crosslinked state. Zhao and coworkers showed for the first time that the incorporated coumarin moieties in block copolymers can stabilize either the core or the shell of micelles [166–168]. Opposite to the cinnamic derivatives, coumarin moieties can undergo a reversible [2+2] photo-cycloaddition upon UV irradiation of the micelle core. The authors demonstrated that the core of the aqueous micellar aggregates can be successfully photo-crosslinked, if the core is prepared from a poly(methacrylate) bearing coumarin side chains. The irradiation wavelength for the crosslinking was above 310 nm, while the reverse process of breaking the bonds occurred below 260 nm. Although, the disruption of the network was not complete, a certain degree of reversibility was achieved. It was also demonstrated that the crosslinked micelles slow down the release of encapsulated molecules compared to the non-crosslinked ones [166, 169].

1.4.1.3 Hydrophobic Chain Fragmentation

Another notable strategy is the incorporation of the *o*-nitrobenzyl group into block copolymers, either as side groups [170, 171], in the main chain [48, 97], or end groups [98]. In a typical example, the *o*-nitrobenzyl side-chain group protects the carboxylic acid in a hydrophobic state; while the hydrophobicity within the protected block drives the micelle formation [171]. Cleavage of the protecting group with UV light irreversibly alters the hydrophobicity of the polymer (now a free carboxylic acid) and disrupts the rearrangement of the micelle leading to cargo release (Fig. 1.7c). In the same way, a coumarin derivative was incorporated in a side-chain fashion to a PEG-b-poly(methacrylic acid) polymer to protect the acid groups [172]. Similar to the *o*-nitrobenzyl derivatives, the coumarin derivative was cleaved from methacrylate using NIR light (795 nm), yielding the free acid. This drastic change in hydrophilicity drove the micelles to disassemble in solution and release their load.

In the main-chain approach, ABA triblock copolymers were prepared with outer blocks consisting in a hydrophilic PEO and the middle block as the photodegradable hydrophobic one [48]. The middle block, prepared by polycondensation, possesses a photocleavable *o*-nitrobenzyl group in each repeating unit so that it can be fragmented into small segments under UV light irradiation. With the micelle formation in aqueous solution, this feature leads to fast photodegradation of the hydrophobic core and thus allows for burst release of loaded species, which is of interest for photo-controlled drug delivery.

An elegant example of end-group fragmentation in micelles was reported by Almutairi group [98]. Polymers with UV and NIR removable end groups that respond to a single light activated event by complete cleavage of the polymer backbone via a self-immolative mechanism were synthesized. Two photocleavable protecting groups were used to end-cap the polymers, namely, *o*-nitrobenzyl alcohol and bromo-coumarin. The polymers were formulated into nanoparticles, and the encapsulated Nile red release upon photolysis of the end groups was investigated. The state of the art of this study resides in the original disassembly process and bursts release of the payload that is only activated on demand. This type of design allows the signal to be amplified in a domino effect to fully degrade the polymer into small molecules, without having to use intense or long periods of irradiation to reach maximal degradation.

1.4.2 Photoresponsive Hydrogels

Hydrogel scaffolds are commonly used as crosslinked carriers for cells because their properties can be tailored to match natural extracellular matrix. For this reason, hydrogels are also used in tissue engineering and regenerative medicine to deliver therapeutic cells to injured or diseased tissue through controlled degradation. The use of erosion of the hydrogel network has been a long-standing method to regulate a compound release. However, a sustained and controlled erosion is necessary to be potentially useful for functional delivery systems, because it could allow for improved site-specific delivery of cargo to diseased tissues, through selective irradiation of the target site. The use of a light stimulus can trigger such an effective release on demand, if the materials are formulated properly. A wide variety of photoresponsive hydrogels have been developed to control material properties; therefore, a selection of most noteworthy examples for drug delivery will be discussed in this section. Photoresponsive hydrogels can exhibit multiple changes upon photoirradiation in their physical and/or chemical properties such as elasticity, viscosity, shape, and degree of swelling and properties which are of interest when modulating the target delivery. To formulate a photoresponsive system, the choice of the photoreactive group and the fundamental structure of the gels are decisive. Both physically (non-covalently) crosslinked and chemically (covalently) crosslinked hydrogel have been developed to incorporate the photoactive group into the hydrogel network. For a better understanding, a comprehensive classification of the most used drug delivery strategies is provided as a schematic illustration. As shown in Fig. 1.8, hydrogel systems can be divided into four categories: (a) chemically irreversible crosslinked hydrogels that incorporate photoresponsive molecules, (b) chemically reversible network-forming hydrogels based on photodimerization, (c) physically reversible network-forming hydrogels containing intermolecular interacting side groups that respond to photoisomerization, and (d)

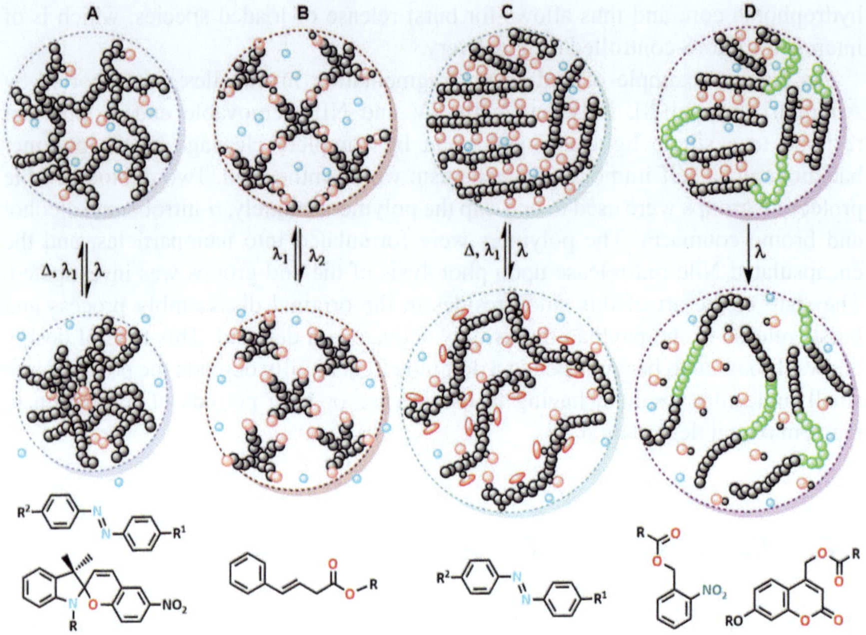

Fig. 1.8 Schematic illustrations for drug release approaches using (**a**) photoinduced swelling–deswelling in the chemically crosslinked networks based on isomerization of photochromic moieties, (**b**) reversible photoinduced crosslinking in a star polymer with photocrosslinkable moieties, (**c**) reversible photoinduced intermolecular assembly from gel to a free-flowing state due to the photoisomerization of the photoresponsive groups, and (**d**) gel network disruption by side-chain fragmentation with removal of photolabile protecting groups

physically irreversible network-forming hydrogels based on the interactions of photocleavable side groups.

1.4.2.1 Photoinduced Swelling–Deswelling in Chemically Crosslinked Hydrogels

In this first category of materials fall the photoresponsive molecular imprinted hydrogels. Molecularly imprinted hydrogels fabricated from tailor-made conformation-switchable functional monomers can alter the strength of the host–guest interaction upon light irradiation [173–179]. A controlled release of the physically bound drugs results when the imprinted binding sites switch under irradiation to a configuration that is not favorable for substrate binding (see Fig. 1.8a). The involvement of photochromic molecules (e.g., azobenzenes or spiropyrans) in the substrate binding sites implies that the release process is reversible, once the transformed binding sites are switched back to their original configuration by a complementary light stimulus. This reversible controlled release and uptake mechanism are fundamentally different from the responsive delivery process of stimuli-induced phase transition materials, where the substrate release process is generally irreversible [180].

An eloquent example of such photoresponsive molecular imprinted hydrogels has been reported by Lam group [173]. The preparation of several hydrogel composition was carried out from a water-soluble azobenzene-containing functional monomer crosslinked with various bisacrylamide and bismethacrylamide bifunctional monomers. The hydrogels were prepared in the presence of paracetamol, chosen as the molecular template for the imprinting. Since photoisomerization of the azobenzene moieties induces structural change of the recognition sites, the release and uptake of paracetamol were controlled by light. This study demonstrated that such photoregulated release and uptake processes are repeatable and can be utilize as a controlled drug delivery system. The same authors reported molecular imprinting hydrogels using caffeine as a molecular template [176]. The photoresponsive molecularly imprinted polymeric (MIP) material was obtained from an azobenzene-based functional monomer and trimethylolpropane trimethacrylate (TRIM) as the crosslinker. Upon irradiation at 365 nm, 58.3% of receptor-bound caffeine is released from the MIP material. Subsequent irradiation at 440 nm causes 96.4% of the released caffeine to be rebound by the MIP material. The near-quantitative uptake of the released caffeine is evidence of the reversibility of the receptor-site configuration and substrate affinity during the photoswitching of the azobenzene chromophores.

Another interesting example of molecularly imprinted hydrogels was reported by Tang's group [177]. A photoresponsive organic–inorganic hybrid material prepared by covalent imprinting of the template–functional monomer complex formed between 4,4′-dihydroxyazobenzene and ibuprofen was developed to explore the photoregulated uptake and release of drug. After imprinting, the template molecules of ibuprofen were removed via acidic hydrolysis, and accurate cavities were left, which could be used as the receptor recognition sites for ibuprofen. The new organic–inorganic hybrid material shows specific affinity to ibuprofen and reversible uptake and release of ibuprofen upon alternate irradiation at 365 and 440 nm, respectively. The results of this work demonstrate the potential of the light-responsive MIP materials as a resourceful approach for drug delivery systems.

1.4.2.2 Photoinduced Reversible Network-Forming Hydrogels

Another feasible approach to confer photoresponsiveness to a hydrogel system is based on reversible crosslinking of photodimerizable moieties. The photoinduced gelation provides several advantages over conventional chemical hydrogel synthesis: mild to moderate reaction conditions, absence of toxic catalysts and initiators, and easily controllable rates of crosslinking (by switching the light source on and off). Unlike small drug molecules, the mesh size of a typical cell-compatible dense hydrogel network in the swollen state is nearly 5 nm. For relatively large protein molecules, adjusting a hydrogel porosity (mesh size) to accommodate fragile biomacromolecules can be engineered to limit mobility and modulate release kinetics using photo-crosslinking [181, 182]. A new generation of hydrogel membranes based on photosensitive four-armed PEG monomers has been developed by Russell group, to adjust the permeation rates of model proteins through the membranes [182]. In this study, the hydroxyl end groups of PEGs were functionalized with cinnamylidene acetate groups to form photosensitive PEG macromers (PEG-CA), which upon irradiation (300 nm) formed crosslinks between adjacent cinnamylidene groups resulting in highly crosslinked networks (hydrogels) (see Fig. 1.8b). The hydrogel membranes were assessed for the permeation fluxes of myoglobin (Mb), hemoglobin (Hb), and lactate dehydrogenase-L (LDH), and the diffusion coefficients of the proteins were determined accordingly. The PEG membranes exhibited a photo-disruptive behavior upon exposure to UV irradiation (254 nm). The diffusion coefficients showed a decreasing trend with the increasing degree of modification of PEG-CA monomers and with the solute molecular size.

1.4.2.3 Photoinduced Reversible Intermolecular Assemblies of Gels

Elegant examples of simple intermolecular assembly based on rheological changes, such as going from a gel to a free-flowing state, were reported [183–185]. These examples fall in the third mentioned category (Fig. 1.8c), and they are meant to show the versatility of azobenzene moieties to influence several hydrogel properties by simply undergoing *trans–cis* photoisomerization. Yu's group reported the preparation of multi-responsive reversible gels based on a carboxylic azo polymer. Due to the specifically designed molecular structure of the carboxylic azo polymer, the polymer would undergo formation of H-aggregation and hydrogen bonding in dimethyl sulfoxide (DMSO). The obtained gel showed multi-response behavior to several stimuli, such as temperature, solvent polarity, and light, and it was systematically studied. It was found that the resulting gels could be disrupted by heating, adding a solvent with lower polarity, and irradiating with UV light. The gel could then be reformed by reverse processes with full reproducibility [183].

1.4.2.4 Photoinduced Irreversible Disruption of Gels by Photolabile Group Removal

Hydrolysis and enzymolysis are the two most common mechanisms employed for hydrogel degradation, but neither allows sequential or staged release of cells. In contrast, photodegradation allows external real-time spatial and temporal control over hydrogel degradation and allows for staged and sequential release of cells. Bettinger and coworkers performed interesting studies on photodegradable physically crosslinked polymer networks (see Fig. 1.8d), having either *o*-nitrobenzyl moieties or coumarin-based blocks incorporated [44, 74]. In the *o*-nitrobenzyl approach, linear triblock copolymers composed of poly (*o*-nitrobenzyl methacrylate) and poly(ethylene glycol) segments of variable molecular weights were synthesized using ATRP [74]. The triblock copolymers self-assemble in gels at room temperature due to physically association of the photolabile side groups. The physically disruption of the 3D networks is induced by photocleavage of the *o*-nitrobenzyl moieties, upon irradiation with long-wave UV light. In the second approach, the physically crosslinked gels can be disrupted through light-induced uncaging of coumarin-based blocks using single- and two-photon absorption irradiation [44]. Both studies demonstrated selective photo-deprotection of self-assembled triblock copolymers. Thus, photodegradable hydrogels exhibit promising applications as biomedical materials that can be manipulated using light.

To advance this elegant proof-of-concept closer toward practical applications, a series of macromers incorporating photodegradable *o*-nitrobenzyl groups in the macromer backbone were developed and characterized by Kasko's group [46, 95]. They formed hydrogels from these macromers via redox polymerization and quantified the apparent rate constants of degradation. Human mesenchymal stem cells (hMSCs) were tested for survive encapsulation in the hydrogels (90% viability post-encapsulation). By exploiting the differences in reactivity of two different *o*-nitrobenzyl linkers, the release of one stem cell population (green fluorescent protein expressing) over another (red fluorescent protein expressing) was demonstrated.

Anseth's group reported another strategy to create photodegradable PEG-based hydrogels, through a rapid polymerization of cytocompatible macromers for remote in situ manipulation of the gel properties [142]. Post-gelation control of the gel properties was demonstrated to introduce temporal changes, creation of arbitrarily shaped features, and on-demand pendant functionality release. This feature allowed for cell migration by active photodegradation of channels within the hydrogel's network. Temporal variation of the biochemical gel composition was used to influence chondrogenic differentiation of encapsulated stem cells.

All expounded example in this subsection were meant to emphasize the potential use in design of photodegradable hydrogels and gels in both drug delivery and tissue engineering.

1.4.3 Photoresponsive Supramolecular Assembly Systems

The introduction of supramolecular chemistry into polymer chemistry provides novel opportunities for the generation of macromolecular architectures with specific functions. Cyclodextrins are attractive design elements as they form supramolecular inclusion complexes with hydrophobic guest molecules in aqueous solution affording the possibility to combine a large variety of building blocks to form novel macromolecular assemblies [186–190]. The ability of polymers modified with cyclodextrins to form gels based on physical interaction with guest molecules like azobenzenes covalently linked to a polymeric matrix (see Fig. 1.9) has been intensively investigated [19, 191–193].

Harada's group formulated a sol–gel system based on azobenzene and α-cyclodextrins [192]. They prepared poly(acrylic acid) bearing azobenzene side

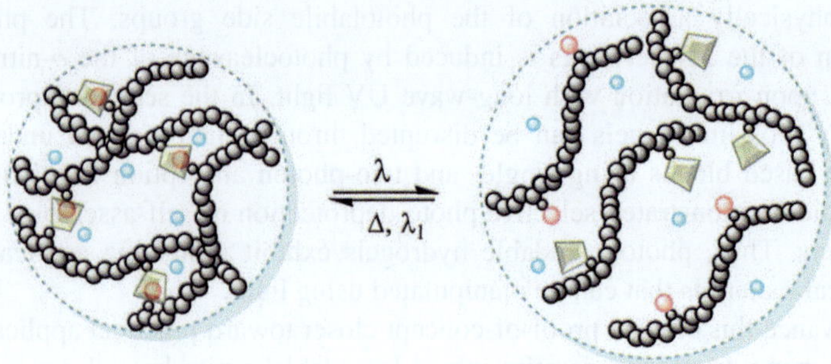

Fig. 1.9 Schematic illustration of drug release approach using a photoinduced reversible supramolecular assembly from gel to a free-flowing state due to the photoisomerization of the photoresponsive groups

groups and glucan curdlan modified with α-cyclodextrin side groups. In this study, the morphology of the supramolecular hydrogels was successfully photoregulated with an appropriate wavelength to control the formation of an inclusion complex. Thus, the complexation between the two polymers controlled by light qualified these supramolecular assemblies for reversible drug delivery systems. To this end, an innovative study of protein release was realized by Kros and coworkers, which used cyclodextrin-modified dextran hydrogels for the in vitro release [143]. Taking advantage of the efficient thiol–maleimide reaction, they functionalized dextrans with either azobenzene (AB) or β-cyclodextrin (CD) moieties. Upon mixing these two polymers, an inclusion of *trans* azobenzene in the β-cyclodextrin induces the gel formation. The inclusion complexes dissociate upon *trans–cis* isomerization of the ABs after irradiation with UV light (365 nm) resulting in a dissolution of the hydrogel. Proteins were physically entrapped in this supramolecular gel matrix simply by dissolving them into the polymer solutions before the gel preparation. Using green fluorescent protein (GFP) as a model protein, the light controlled in vitro release was demonstrated. The current characteristics of this system can serve for the future drug screening technology using zebrafish embryos, due to the transparency of these systems for the 365 nm UV light.

Another particular example of supramolecular assembly that showed potential for drug release was reported by Yuan's group [194]. They prepared light-controlled polymeric nanotubes based on terminal host–guest interactions of two linear homopolymers. Reversible assembly and dissociation of poly(ε-caprolactone) (PCL)-cyclodextrin/poly(acrylic acid) (PAA)-*trans*-azobenzene nanotubes are achieved by alternating irradiation with UV and visible light. In this study, nearly 100% release of an encapsulated model payload composed of rhodamine B is reported. Considering the active nature of the host–guest interactions that enables self-assembly at a higher order makes this complex system a promising carrier for light-controlled drug release.

Based on the existing productive reports, it is anticipated that this line of work may open a pathway for fabricating new polymeric containers which can be used as functional photo-controllable drug delivery systems.

1.5 Bioimaging

Bioimaging relates to methods that noninvasively visualize biological processes in real time. In vitro or in vivo bioimaging techniques are valuable tools for clinical diagnostics. Different types of bioimaging methods are available, ranging from techniques that enable organism anatomical imaging (i.e., magnetic resonance imaging, MRI) to others that provide specific molecular imaging (i.e., optical fluorescence) at subcellular resolution. Such tools are expected to be pivotal for advancing early-stage cancer diagnosis, guided stem cell therapies, drug delivery, pathogen detection, gene therapy, image-guided surgery, and cancer staging, in addition to many other clinically relevant procedures, diagnostics, and therapies [195].

The authors acknowledge that bioimaging is a wide field of research comprising many types of techniques and chemical compounds. Our focus will be on including the light-responsive materials belonging to azobenzenes, coumarin, and spiropyran classes that were reported in this type of application, highlighting the advantages and disadvantages compared with other fluorophores. For other types of fluorophores and more exhaustive studies, the reader should consult the following reviews [195–197].

The fluorescence microscopy has revolutionized the visualization of structures and dynamics in biology. It proved to be a very useful technique and its uses lead to major discoveries. Fluorescence and autofluorescence, thoroscopic and endoscopic techniques are emerging as powerful diagnostic tools for identifying disease and abnormal structural features on body cavity surfaces. When imaging technique is noninvasive and used on thin tissue sections, then the fluorescence in visible region is the appropriate choice. If deeper penetration depth is needed especially for clinical applications, then fluorescence-based techniques into the NIR region must be used (650–1000 nm).

1.5.1 Organic Photoresponsive Probes

Applications of *azobenzene* in bioimaging are very limited mainly because their photoresponse can be efficiently controlled only using UV light. Although azobenzenes that respond to green light like Dispersed Red 1 and others have been reported, they lack thermal stability of the *cis* isomers. Exhibiting very fast thermal relaxation, milliseconds to seconds, these azo-derivatives become unsuitable for imaging application where longer timescale is needed. Another important issue is the toxicity and stability of these azo compounds in physiological media. Still efforts to overcome these disadvantages are being made.

Using an innovative strategy, Woolley et al. demonstrated the utility of azobenzene photoswitching in in vivo fluorescence imaging. The experiments were done on zebrafish by using fluorescent peptides conjugated to azobenzene, so that the fluorescence becomes apparent upon photoactivating the molecule with light of appropriate wavelength [198].

New sulforhodamine-based fluorescent "turn-on" probes have been developed for the direct imaging of cellular hypoxia by Chevalier et al. [199]. Rapid access to this novel class of water-soluble "azobenzene-caged" fluorophores was made possible through an easily implementable azo-coupling reaction between a fluorescent primary arylamine derived from a sulforhodamine 101 scaffold and a tertiary aniline whose N-substituents are neutral, cationic, or zwitterionic. The probe was successfully applied for imaging the hypoxic status of tumor cells (A549 cells).

Spiropyrans have two main advantages in comparison with azobenzenes in respect to bioimaging applications: they have fluorescence and they can be switched between the two stable states by changing the pH of the medium. Because of their facile synthesis, distinct photochromic transition, and high fatigue resistance, the spiropyran chromophores have been thoroughly investigated.

Marriott et al. introduced a Förster resonance energy transfer (FRET) imaging method using an optical switch probe, nitrobenzospiropyran (NitroBIPS), which substantially improved the sensitivity of detection to <1% FRET efficiency [200]. Through orthogonal optical control of the colorful merocyanine and colorless spiro states of the NitroBIPS acceptor, donor fluorescence could be measured both in the absence and presence of FRET in the same FRET pair in the same cell. In vivo imaging studies of a green fluorescent protein-alkylguaninetransferase (GFP-AGT) NitroBIPS complex, employing optical lock-in detection of FRET, allowed unambiguous resolution of FRET efficiencies below 1%, equivalent to a few percent of donor-tagged proteins in complexes with acceptor-tagged proteins.

Sensor for extreme pH values, useful in detection of microorganisms including acidophile, *Helicobacter pylori*, *Escherichia coli* (*E. coli*), and *Bacillus subtilis* (*B. subtilis*) were developed by Yin et al. [201]. The spiropyran modified with bulky benzoic acid group at the indole N-position and the electron withdrawing carboxyl group at the indole 6-position have shown excellent extreme pH sensor properties due to its two-color fluorescence, i.e., red in extreme acid, blue-green in extreme alkali, and nonfluorescence at intermediate pH. In vivo experiments showed that the dye efficiently labeled the bacterial cells at extreme pH with the same effect as in solution. Another advantage of this sensor was that it could be trigger by visible-NIR light.

The *coumarin* derivatives have found multiple applications in sensing area and live cell imaging due to their desirable photophysical properties, such as a large Stoke's shift, visible emission wavelengths, and high fluorescence quantum yields. Application of coumarin in bioimaging field is based on their fluorescence properties and not on their photodimerization reaction. Transition metals play a very important role in the human body, the gene transcription, regulation of metalloenzymes, neural signal transmission, and apoptosis. Therefore, quantitative measurement of metals traces in vivo is a very important area of research. Because of the lack of spectroscopic signature of some metal ions, the use of fluorescence probes has become essential in cell imaging. Detection of several metals like Zn^{2+}, Cu^{2+}, and Al^{3+} was possible by using rational designed coumarin dyes [202–204].

Fluorescent chemosensors based on coumarine dyes for detection of metals in living cells were developed by Zeng et al. [204]. The sensors displayed high selectivity toward Cu^{2+} ions and high sensitivity. The experiments were carried out using liver cell lines in which Cu^{2+} is known to be accumulated. One interesting approach in metal ion detection was the synthesis of cleavable coumarin-based probes [205]. The hydrolysis reaction generates 3-amino-7-hydroxycoumarin and 3,4-dihydroxybenzaldehyde, where 3-amino-7-hydroxycoumarin is the actual substance reacting with Cu^{2+} or Fe^{3+} ions and undergoing fluorescence quenching. Colorimetric and ratiometric fluorescent probe for sulfite with coumarin as the fluorophore and α,β-unsaturated ketone as the receptor were reported [206]. High selectivity and good competition of the probe toward sulfite over other anions and thiols enable to monitor sulfite levels in realistic samples as well as in living HeLa cells. Coumarin derivatives have been used in more exotic applications like Au^{3+} detection in living cells [207]. The starting julolidine acetylenic chromophores were very weekly fluorescent. By adding catalytic amounts of gold, mediated hydroarylation reaction takes place, and a strongly fluorescent coumarin derivative is formed. The probes identified Au^{3+} ions highly selectively over other biologically relevant metal ions, and detection concentration of Au^{3+} was as low as 0.1 μM. Coumarin derivatives have found applications also in bioimaging of cyanide anions [208], which are very important in human health area, and nitroxyl (HNO), a reactive nitrogen specie, which mediates a variety of biological events, related especially to heart problems [209].

Detection of intracellular thiols, such as cysteine, homocysteine, or glutathione, is another important area where coumarin probes were used [210, 211]. A colorimetric and turn-on fluorescent probe based on a coumarin unit for rapid and highly selective GSH detection in PBS buffer solution (pH = 7.4) was designed and synthesized by Liu et al. [212]. Moreover, they successfully carried out fluorescence imaging of HeLa and Hek-293a cells, to prove the cell permeability and biocompatibility of the probe. A series of crescent-shaped, deep-red emission dyes that contain coumarin- and pyronin-fused conjugated skeletons (CP dyes) were designed and applied in fluorescence imaging experiments [213]. The hybridization of coumarin and pyronin Y moieties induces significant stock shifts and long-wavelength emissions. Using live cell fluorescence imaging experiments the authors proved that different amino substituents significantly influence the imaging properties of CP dyes, including cell membrane permeability, nuclear membrane permeability, and the organelle location. They concluded also that the low cell cytotoxicity and high photostability make CP dyes good options for monitoring the dynamic changes in mitochondria, lysosomes, or nucleoli in living cells.

Fluorescence microscopy has one major drawback: the resolution limit of about 250 nm laterally and 500 nm axially, such that features closer than these distances are obscured. This resolution limit is caused by the diffraction of light and is a characteristic of all far-field light microscopies. Other biological imaging modalities, such as atomic force microscopy, electron microscopy, or near-field techniques, can attain very high spatial resolution but either have a poor ability to identify the molecules being examined or are primarily limited to the imaging of surface features [214]. Recently, super-resolution microscopy has emerged as a promising imaging tool that can overcome the light diffraction barrier and obtain high spatial resolution. With the emergence of these techniques, the light-responsive materials have also started to receive considerable interest.

Specially designed chromophores belonging to coumarin and spiropyran classes have found to be suitable for this type of microscopy. Hell and coworkers synthesized a series of coumarin-based derivatives, which are photostable and highly fluorescent and exhibit large Stokes shifts [215]. They have also decorated the coumarin dye with phosphoric group to enhance the hydrophilicity and to have low self-aggregation. The new coumarin, with red fluorescence emission, was further used in multicolor confocal and stimulated emission depletion microscopy for the sharp, distinct localization of the *cis*-Golgi and *trans*-Golgi networks in the Golgi apparatus.

Zhang and coworkers employed spiropyran dyes to design a super-resolution imaging sensor for the quantitative identification of tubulin in the microtubules of live cancer cells [216]. They successfully demonstrated the site-selective detection of tubulin with a spatial resolution of approximately 20 nm. Photoactivatable fluorophores for super-resolution imaging based on oxazine auxochromes were synthesized by Raymo [217]. These compounds pair an oxazine photochrome to a biphenyl, fluorene, pyrene, coumarin, or cyanine fluorophore in their molecular skeleton and could be prepared in a single step from known precursors. The brightness of the compound modified with a coumarin group was sufficiently high to permit the imaging of individual molecules with excellent signal-to-noise ratios.

1.5.2 Macromolecular Photoresponsive Probes

Embedding light-responsive compounds into polymer matrix offers several advantages in bioimaging applications like thermal stability, mechanical strength, lower toxicity, and higher detention time. Light-responsive polymeric materials can be obtained by doping the polymer with light-responsive dyes or by covalently attaching the dye. Each method has advantages and disadvantages, but the most important is that by chemical linkage, no dye leakage can take place. A temperature-responsive block copolymer with excellent hydrophilicity and biocompatibility based on N-isopropylacrylamide and 7-amino-4-methylcoumarin was reported by Chen et al. [218]. They proved that the copolymer can be used in the field of high-resolution imaging of temperature profiles, based on the variation of fluorescence intensity with temperature. Dual stimuli-responsive (pH and light) hyperbranched star copolymers based on a sugar derivative and spiropyran modified with polymerizable bond were successfully applied in cell imaging [219]. The copolymers were prepared by two-step RAFT polymerization: first the hyperbranched polymer, poly(6-O-methacryloyl-1,2,3,4-diO-isopropylidene-D-galactopyranose) was prepared and then was used as the macro-RAFT agent in the subsequent RAFT copolymerization of 2-(N,N-dimethylaminoethyl) methacrylate and 1′-(2-methacry-loxyethyl)3′,3′-dimethyl-6-nitro-spiro(2H-1-benzo-pyran-2,2′-indoline).

1.5.3 Photoresponsive Polymeric Nanoparticles

Photoresponsive organic molecules suffer from inferior photophysical properties (photo bleaching and brightness), which can be improved by covalently integrating them into polymeric scaffolds [110, 220]. Dye-doped polymer nanoparticles (NPs) are crosslinked polymeric spheres in which dyes were covalently attached or physically entrapped. Light-responsive polymeric nanoparticles can be synthesized through copolymerization of light-responsive monomers with other functional monomers [221, 222] or by the encapsulation of organic dye molecules inside or on the matrix [42, 223]. Encapsulation of dyes into nanoparticles provides advantages, over the use of dyes alone: improved contrast, reduced chemical and photodegradation, signal amplification due to increase number of molecules in the nanoparticles, and the ability to target specific biologically relevant sites.

Using spiropyran-based photoswitchable polymeric NPs, Li and coworkers developed the PULSAR microscopy technique and achieved biological fluorescence imaging with resolutions down to 10–40 nm [224]. The greatest advantage of this approach was that spiropyran-to-merocyanine photoswitching imparts strong red fluorescence, and the resulting merocyanine offers good photophysical characteristics including a high emission rate, long photobleaching time, and large total detected photon count for single-molecule imaging. The same research group developed a class of spiropyran dyes, and their fluorescence colors could be reversibly photoswitched from red to green, blue, or nearly dark, thus alternating between two colors [225]. The nanoparticles incorporating these photoswitchable dyes enabled to emit either one pure color or a combination of both colors because the nanoparticle fluorescence originates from multiple dyes. Thus, the nanoparticle shines >30 times brighter than state-of-the-art organic dyes such as fluorescein. Decorated with HMGA1 protein, these optically switchable dual-color nanoparticles undergo endocytosis and unambiguously identify themselves from fluorescence interference including autofluorescence, thus enabling a new tool for live cell imaging.

Dual-color fluorescent tags, especially those that can be reversibly photoswitched, are particularly desirable because they can distinguish sites of interest from false-positive signals generated by adventitious fluorescent biomolecules. Optically switchable dual-color fluorescent nanoparticles that incorporate two classes of dyes

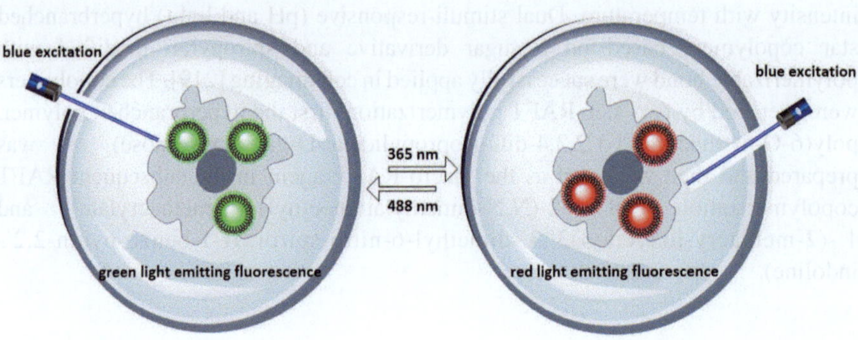

Fig. 1.10 Light emission of dual-color fluorescent nanoparticles. When the photoswitchable dye is in its ring-closed (spiro) form, the perylene chromophore located in the hydrophobic core strongly emits a distinctive green fluorescence. When UV-induced ring opening converts the spiro compound to its open-ring form (merocyanine), whose visible absorption band nearly coincides with the perylene emission bands, the perylene emission is strongly quenched by fluorescence resonance energy transfer (FRET). Consequently, in the hydrophobic cavities of the water-soluble nanoparticles, the mero form acquires strong red fluorescence (600–750 nm, with $\Phi = 0.18$ estimated by comparison to a rhodamine B standard)

into the polymeric chains have been reported in the literature by using an emulsion polymerization method [81]. The nanoparticles consist of an organic photoisomerizable dye, spiropyran, as an optically responsive component, and another fluorescent dye, perylene diimide, as a high-energy emitter. These dual-color fluorescent nanoparticles were developed as potential new tools for biomedical applications and live cell imaging. When delivered into HEK-293 cells, they display either red or green fluorescence, depending upon the wavelength of light to which the cells were highlighted (see Fig. 1.10).

The probe brightness is of great importance in biological fluorescence imaging in terms of elucidating the signals of interest from complicated physiological environments with high levels of background noise. Single probe has limited brightness; therefore, nanoparticles containing probes can represent a viable solution of this problem. Taking the advantage of fluorescence photoswitching features of the spiropyran-containing polymeric NPs, Li and coworkers developed photoswitching-enabled Fourier transform (PFT) fluorescence microscopy [226]. The biggest advantage of this technique is that by measuring more oscillating cycles, they can continuously amplify the brightness of the modulated probes.

Nanoparticles doped with specially designed oxazine dyes modified with coumarin units were successfully applied in super-resolution imaging and demonstrated an excellent blinking effect at the single-molecule level with a spatial resolution of 28 nm [217].

1.6 Analytical Techniques and Characterization of Photoresponsive Polymers

The continuing expansion in the field of photoresponsive polymers has led to the development of new techniques of characterization. Hereinafter, we will try to provide a short description of the state-of-the-art techniques used to characterize photoresponsive polymers with respect to biomedical applications. The techniques based on their optical properties and the ones characterizing the surface properties, which are crucial for biomedical applications, will be highlighted.

Two-photon fluorescence microscopy was discovered in 1990 by Webb [227], and since then its impact in bioimaging domain has increased dramatically. In this technique, the fluorescence is induced by two-photon excitation, which is a great advantage because it provides higher spatial resolution using excitation in the near-infrared spectral window (i.e., the biological materials have the lowest absorption in this region). Also, other important advantages are reduced photobleaching and photodamage of biological sample and high-contrast three-dimensional images. Important progress in visualization of different organs like the kidney [228], brain [229], eye [230], and cardiovascular tissue [231] or detection of malignancy in organs has been realized [232]. In two-photon microscopy, two photons of approximately twice the wavelength of single-photon excitation are used to excite the fluorophore to the excited state. Emission process happens in the same way as with the one-photon excitation. The most important parameters to be determined for two-photon fluorophores are the two-photon action cross section, the fluorescence quantum yield, two-photon absorption cross section, and the photostability. The details of this method as well as the chromophores used have been well documented and reviewed in numerous publications [233–237].

Super-resolution microscopy is another field where photoresponsive molecules have found important applications [238, 239]. The work of Prof. Stefan W. Hell, Prof. Eric Betzig, and Prof. William Esco Moerner in this field was rewarded with the prestigious Nobel Prize in 2014. Super-resolution imaging is a sort of far-field optical microscopy technique, in which the separation between the lens and the sample is much larger than the light wavelength, compared with a near-field scanning optical microscope.

Various innovative techniques have been reported up to now in this field. One of the most important is stimulated emission depletion microscopy (STED) which was first introduced by Hell [240, 241]. In conventional microscopy, the fluorescence emission resulting from an excitation laser beam by spontaneously excited fluorophores leads to the formation of a diffraction-limited image. In STED microscopy, a second laser is used to suppress the fluorescence emission from the fluorophores located off the center of the excitation. When an excited-state fluorophore encounters a photon that matches the energy difference between the excited and the ground state, it can be brought back to the ground state through stimulated emission before spontaneous fluorescence emission occurs. This process effectively depletes excited-state fluorophores capable of fluorescence emission.

Fluorophores specially designed belonging to coumarin and spiropyran classes have been used with great success in this type of microscopy [238].

Photo-activated localization microscopy (PALM) and stochastic optical reconstruction microscopy (STORM) are based on the same principle: stochastic switching of individual fluorophores between the *on* state (fluorescent) and the *off* state (nonfluorescent). The stochastic photoswitching of the fluorophores provides rare fluorescent spots in each frame of image, which facilitates the accurate localization of a single fluorophore after an analysis algorithm. In this technique, rhodamine-, cyanine-, and spiropyran-type chromophores have found applications [238].

However, these super-resolution techniques are challenging to use in live tissues, for two reasons. First, the living system is dynamic and these techniques require longer times to acquire the data. Second, the imaging depth in thick tissue is still limited to less than 100 μm, and the scattering limits the resolution improvements very rapidly as depth is increased [237].

The surface structure characterization of a biomaterial is very important, since the surface is the interface where the biomaterial meets and interacts with the molecular constituents of the biological host. These interactions are crucial for the successful performance of the material in the biological media. For investigating physical properties of the material surface, techniques like atomic force microscopy (AFM), scanning electron microscopy (SEM), transmission electron spectroscopy (TEM), scanning tunneling microscopy (STM), and nanoindentation are being used.

SEM, TEM, and AFM can be categorized as daily basis characterization methods for photoresponsive biomaterials. However, confocal laser scanning microscopy (CLSM) is a relatively new method used for obtaining high-resolution images and 3-D reconstructions of a variety of biomaterial samples. The principle of this method is based on a laser beam of light that is focused onto the specimen by a microscope objective lens. The backscattered light signal or fluorescence signal from the sample is then collected by the same microscope objective and recorded by a photodetector. By stepping the focus vertically (z), multiple slices can be used to build up a full three-dimensional image. Zhao et al. synthesized light-responsive nanocarriers, and using CLSM they were able to visualize the cellular uptake and intracellular doxorubicin release in HeLa cancer cells [242]. CLSM was also used to highlight that folic acid conjugated coumarin photoresponsive polymers target efficiently the tumor cells through folate receptor-mediated endocytosis [68].

Nanoindentation is another supportive characterization technique that can be used to determine thin films properties and also in the study of soft, hydrated biomaterials such as hydrogels [243]. The principle of nanoindentation is very simple. The indenter tip with a known geometry is driven into a specific site of the material to be tested, by applying an increasing normal load. When reaching a preset maximum value, the normal load is reduced until complete relaxation occurs. The load is applied by a piezo actuator, and the load is measured in a controlled loop with a high sensitivity load cell. The resulting load/displacement curves provide data specific to the mechanical nature of the material under examination. Established models are used to calculate quantitative hardness and modulus values for such data. Photomechanical polymers are characterized by changes in their mechanical properties under light radiation. This process is driven by the reversible isomerization of chromophores in the polymer matrix; therefore, nanoindentation is a very useful technique to test the stiffness of photoresponsive biomaterials [244–246].

1.7 Conclusions and Outlook

Though exciting progress has been made in photoresponsive polymers, thanks to development and advances in controlled radical polymerizations that have contributed to the engineering of advanced multifunctional polymeric structures with precise control over architecture, these materials are yet to reach the practical realm. From a fundamental perspective, while significant information has been obtained about the initial and final states of these systems in response to light, still major drawbacks arise from the limitations of photophysical properties of the photoactive species.

Although for azobenzenes there is already a large body of literature that describes the influence of the syntheses approach and the structural features on their optical properties, together with the toxicity of photoswitches, and the products of their degradation, these compounds are not self-sufficient to bring the photoresponsive systems they compose of, to the "clinic" stage. The major shortcoming of these photoactive compounds represents the narrow window of red-shifted azobenzene derivatives. Moving the absorption bands of photoswitchable azo-derivatives into the optical window between $\lambda = 650$ nm and 900 nm will offer attractive alternatives to apply near-IR irradiation that can penetrate deeper. Also, azo-derivatives with more stable *cis* isomers will be beneficial in some domains like photopharmacology. The UV absorbing photoactive compounds such as cinnamic acid or coumarin that undergo photodimerization exhibit the same undesirable trait. However, these compounds have found their utility in domains other than the medical field, due to their ability to reversibly alter polymer properties, such as flexing, bending, and healing.

Unlike the azobenzenes, the spiropyrans and spirooxazines have a favorable photoswitching window, good resistance to fatigue, and stable switchable isomer species; however, they lack chemical diversity; therefore, they exhibit rather limited photophysical properties. Despite these shortcomings, azobenzene, spiropyrans, and other photochromic molecules remain valuable photoactive compounds of choice due to their "clean" photoreactions and the optical reversibility they confer to polymeric systems they incorporate.

The *o*-nitrobenzyl and coumarin photocleavable derivatives have been widely investigated as well for their potential use in the biomedical field. The most noteworthy feature of these derivatives is that they can be cleaved photochemically with near-IR light by two-photon absorption processes to alter polymeric system's properties on demand. The versatility in synthesis is also a trait that favors wide architectural compositions. However, they lack the chemical reversibility of the photoisomerization processes, and some of the photocleaved small molecules that are released in the process are often toxic.

Light-responsive molecules are indispensable components in bioimaging. Azobenzene compounds seem to have many limitations (low fluorescence yield and biocompatibility and weak photoregulation with UV light) regarding their applications in bioimaging, especially for in vivo applications. Still they could represent a viable solution for in vitro measurements like live cell imaging where UV light can be used. Spiropyran and coumarin derivatives have proved their usefulness, especially in the field of super-resolution microscopy, but still steps must be made in order to be used for in vivo imaging.

Despite impressive progress, many challenges remain. It remains difficult to achieve high local densities of dyes without undesired intermolecular interactions, which self-quench fluorescence and decrease energy transfer efficiency. Using polymeric nanoparticles seems to be the rightful choice to tackle these problems. One can speculate that the continued progress in bioimaging field is focused toward learning how to exploit more efficiently various interactions between polymeric substrate and dyes and to fabricate new supramolecular assemblies with tunable optical properties.

The wide diversity of photoactive compounds has opened pathways to new possibilities in design that were previously unattainable and have generated plenty useful studies for in vitro routine. With inspirations arising from a variety of sources, novel polymers with sensitivities to light as trigger are constantly being developed. So far it was proven to be chemically impossible to synthesize *the perfect photoactive compound*, which combines all the positive features and none of the negative ones. For applications like drug delivery and diagnostics, materials with the ability to respond to several stimuli would be extremely beneficial. The field of photoresponsive polymers will continue to develop, and with it the ability to prepare macromolecules with topological complexity will also expand. Combination of light and other stimuli will take on greater roles in the next generation of stimuli-responsive materials.

References

1. P. Schattling, F.D. Jochum, P. Theato, Multi-stimuli responsive polymers—the all-in-one talents. Polym. Chem. **5**, 25–36 (2014)
2. R. Langer, D.A. Tirrell, Designing materials for biology and medicine. Nature **428**, 487–492 (2004)
3. M.C. Stuart, W.T.S. Huck, J. Genzer, et al., Emerging applications of stimuli-responsive polymer materials. Nat. Mater. **9**, 101–113 (2010)
4. Y. Zhao, Light-responsive block copolymer micelles. Macromolecules **45**, 3647–3657 (2012)
5. J.-M. Schumers, C.-A. Fustin, J.-F. Gohy, Light-responsive block copolymers. Macromol. Rapid Commun. **31**, 1588–1607 (2010)

6. J.S. Katz, J.A. Burdick, Light-responsive biomaterials: development and applications. Macromol. Biosci. **10**, 339–348 (2010)
7. A. Goulet-Hanssens, C.J. Barrett, Photo-control of biological systems with azobenzene polymers. J. Polym. Sci. A Polym. Chem. **51**, 3058–3070 (2013)
8. F. Ercole, T.P. Davis, R.A. Evans, Photo-responsive systems and biomaterials: photochromic polymers, light-triggered self-assembly, surface modification, fluorescence modulation and beyond. Polym. Chem. **1**, 37–54 (2010)
9. T. Ikeda, Photomodulation of liquid crystal orientations for photonic applications. J. Mater. Chem. **13**, 2037–2057 (2003)
10. M.C. Spiridon, K. Iliopoulos, F.A. Jerca, et al., Novel pendant azobenzene/polymer systems for second harmonic generation and optical data storage. Dyes Pigments **114**, 24–32 (2015)
11. J.A. Delaire, K. Nakatani, Linear and nonlinear optical properties of photochromic molecules and materials. Chem. Rev. **100**, 1817–1845 (2000)
12. F.D. Jochum, P. Theato, Temperature- and light-responsive smart polymer materials. Chem. Soc. Rev. **42**, 7468–7483 (2013)
13. H. Tian, Z. Tang, X. Zhuang, et al., Biodegradable synthetic polymers: preparation, functionalization and biomedical application. Prog. Polym. Sci. **37**, 237–280 (2012)
14. G. Pasparakis, T. Manouras, P. Argitis, et al., Photodegradable polymers for biotechnological applications. Macromol. Rapid Commun. **33**, 183–198 (2012)
15. B.H. Cumpston, S.P. Ananthavel, S. Barlow, et al., Two-photon polymerization initiators for three-dimensional optical data storage and microfabrication. Nature **398**, 51–54 (1999)
16. H.Y. Jiang, S. Kelch, A. Lendlein, Polymers move in response to light. Adv. Mater. **18**, 1471–1475 (2006)
17. A. Lendlein, H.Y. Jiang, O. Junger, et al., Light-induced shape-memory polymers. Nature **434**, 879–882 (2005)
18. T. Ikeda, M. Nakano, Y.L. Yu, et al., Anisotropic bending and unbending behavior of azobenzene liquid-crystalline gels by light exposure. Adv. Mater. **15**, 201 (2003)
19. Y.-L. Zhao, J.F. Stoddart, Azobenzene-based light-responsive hydrogel system. Langmuir **25**, 8442–8446 (2009)
20. K. Han, W. Su, M. Zhong, et al., Reversible photocontrolled swelling-shrinking behavior of micron vesicles self-assembled from azopyridine-containing diblock copolymer. Macromol. Rapid Commun. **29**, 1866–1870 (2008)
21. S. Xie, A. Natansohn, P. Rochon, Recent developments in aromatic azo polymers research. Chem. Mater. **5**, 403–411 (1993)
22. T.M. Geue, A.G. Saphiannikova, O. Henneberg, et al., Formation mechanism and dynamics in polymer surface gratings. Phys. Rev. E **65** (2002)
23. T. Ubukata, T. Seki, K. Ichimura, Surface relief gratings in host-guest supramolecular materials. Adv. Mater. **12**, 1675 (2000)
24. V. Shibaev, A. Bobrovsky, N. Boiko, Photoactive liquid crystalline polymer systems with light-controllable structure and optical properties. Prog. Polym. Sci. **28**, 729–836 (2003)
25. S. Kurihara, T. Ikeda, S. Tazuke, et al., Isothermal phase-transition of liquid-crystals induced by photoisomerization of doped spiropyrans. J. Chem. Soc.-Faraday Trans. **87**, 3251–3254 (1991)
26. A.S. Angeloni, D. Caretti, C. Carlini, et al., Photochromic liquid-crystalline polymers main chain and side-chain polymers containing azobenzene mesogens. Liq. Cryst. **4**, 513–527 (1989)
27. Z. Yue, I. Tomiki, *Smart light-responsive materials: azobenzene-containing polymers and liquid crystals* (John Wiley & Sons, Inc., Hoboken, NJ, 2009)
28. T. Seki, M. Sakuragi, Y. Kawanishi, et al., Modulated photoregulation of liquid-crystal alignment by azobenzene Langmuir-Blodgett layers—reversible alignment changes of liquid-crystals induced by photochromic molecular films 0.11. Thin Solid Films **210**, 836–838 (1992)

29. A. Natansohn, P. Rochon, J. Gosselin, et al., Azo polymers for reversible optical storage 0.1. poly 4′- 2-(acryloyloxy)ethyl ethylamino-4-nitroazobenzene. Macromolecules **25**, 2268–2273 (1992)
30. S. Kawata, Y. Kawata, Three-dimensional optical data storage using photochromic materials. Chem. Rev. **100**, 1777–1788 (2000)
31. F.M. Andreopoulos, I. Persaud, Delivery of basic fibroblast growth factor (bFGF) from photoresponsive hydrogel scaffolds. Biomaterials **27**, 2468–2476 (2006)
32. J. Edahiro, K. Sumaru, Y. Tada, et al., In situ control of cell adhesion using photoresponsive culture surface. Biomacromolecules **6**, 970–974 (2005)
33. A. Higuchi, A. Hamamura, Y. Shindo, et al., Photon-modulated changes of cell attachments on poly(spiropyran-co-methyl methacrylate) membranes. Biomacromolecules **5**, 1770–1774 (2004)
34. C. Zhu, C. Ninh, C.J. Bettinger, Photoreconfigurable polymers for biomedical applications: chemistry and macromolecular engineering. Biomacromolecules **15**, 3474–3494 (2014)
35. I. Tomatsu, K. Peng, A. Kros, Photoresponsive hydrogels for biomedical applications. Adv. Drug Deliv. Rev. **63**, 1257–1266 (2011)
36. H. Priya James, R. John, A. Alex, et al., Smart polymers for the controlled delivery of drugs—a concise overview. Acta Pharm. Sin. B **4**, 120–127 (2014)
37. R. Cheng, F. Meng, C. Deng, et al., Dual and multi-stimuli responsive polymeric nanoparticles for programmed site-specific drug delivery. Biomaterials **34**, 3647–3657 (2013)
38. A.K. Bajpai, S.K. Shukla, S. Bhanu, et al., Responsive polymers in controlled drug delivery. Prog. Polym. Sci. **33**, 1088–1118 (2008)
39. M. Behl, M.Y. Razzaq, A. Lendlein, Multifunctional shape-memory polymers. Adv. Mater. **22**, 3388–3410 (2010)
40. L. Zhai, Stimuli-responsive polymer films. Chem. Soc. Rev. **42**, 7148–7160 (2013)
41. M. Martina, D.W. Hutmacher, Biodegradable polymers applied in tissue engineering research: a review. Polym. Int. **56**, 145–157 (2007)
42. E. Fleige, M.A. Quadir, R. Haag, Stimuli-responsive polymeric nanocarriers for the controlled transport of active compounds: concepts and applications. Adv. Drug Deliv. Rev. **64**, 866–884 (2012)
43. H. Nishioka, X. Liang, T. Kato, et al., A photon-fueled DNA nanodevice that contains two different photoswitches. Angew. Chem. Int. Ed. **51**, 1165–1168 (2012)
44. C. Zhu, C.J. Bettinger, Light-induced remodeling of physically crosslinked hydrogels using near-IR wavelengths. J. Mater. Chem. B **2**, 1613–1618 (2014)
45. A. Serafim, C. Tucureanu, D.G. Petre, et al., One-pot synthesis of superabsorbent hybrid hydrogels based on methacrylamide gelatin and polyacrylamide. Effortless control of hydrogel properties through composition design. New J. Chem. **38**, 3112–3126 (2014)
46. D.R. Griffin, A.M. Kasko, Photodegradable macromers and hydrogels for live cell encapsulation and release. J. Am. Chem. Soc. **134**, 13103–13107 (2012)
47. G. Liu, W. Liu, C.-M. Dong, UV- and NIR-responsive polymeric nanomedicines for on-demand drug delivery. Polym. Chem. **4**, 3431–3443 (2013)
48. D. Han, X. Tong, Y. Zhao, Fast photodegradable block copolymer micelles for burst release. Macromolecules **44**, 437–439 (2011)
49. Q. Jin, G. Liu, J. Ji, Micelles and reverse micelles with a photo and thermo double-responsive block copolymer. J. Polym. Sci. Part A: Polym. Chem. **48**, 2855–2861 (2010)
50. J. Jiang, Q. Shu, X. Chen, et al., Photoinduced morphology switching of polymer nanoaggregates in aqueous solution. Langmuir **26**, 14247–14254 (2010)
51. E. Cabane, V. Malinova, W. Meier, Synthesis of photocleavable amphiphilic block copolymers: toward the design of photosensitive nanocarriers. Macromol. Chem. Phys. **211**, 1847–1856 (2010)
52. Y. Zhao, Photocontrollable block copolymer micelles: what can we control? J. Mater. Chem. **19**, 4887–4895 (2009)

53. M.W. Urban, Stratification, stimuli-responsiveness, self-healing, and signaling in polymer networks. Prog. Polym. Sci. **34**, 679–687 (2009)
54. Y. Hirshberg, Reversible formation and eradication of colors by irradiation at low temperatures. A photochemical memory model. J. Am. Chem. Soc. **78**, 2304–2312 (1956)
55. E. Kim, Photorefractive effects in organic photochromic materials, in *Photorefractive materials and their applications* 2, ed. by W. T. Rhodes, (Springer Science+Business Media, LLC, New York, 2007), p. 607
56. F.A. Jerca, V.V. Jerca, D.F. Anghel, et al., Novel aspects regarding the photochemistry of azo-derivatives substituted with strong acceptor groups. J. Phys. Chem. C **119**, 10538–10549 (2015)
57. Z. Mahimwalla, K.G. Yager, J.-I. Mamiya, et al., Azobenzene photomechanics: prospects and potential applications. Polym. Bull. **69**, 967–1006 (2012)
58. A.A. Beharry, G.A. Woolley, Azobenzene photoswitches for biomolecules. Chem. Soc. Rev. **40**, 4422–4437 (2011)
59. A.A. Beharry, L. Wong, V. Tropepe, et al., Fluorescence imaging of azobenzene photoswitching in vivo. Angew. Chem. Int. Ed. **50**, 1325–1327 (2011)
60. A.A. Beharry, O. Sadovski, G.A. Woolley, Azobenzene photoswitching without ultraviolet light. J. Am. Chem. Soc. **133**, 19684–19687 (2011)
61. L. Florea, D. Diamond, F. Benito-Lopez, Photo-responsive polymeric structures based on spiropyran. Macromol. Mater. Eng. **297**, 1148–1159 (2012)
62. G. Such, R.A. Evans, L.H. Yee, et al., Factors influencing photochromism of spiro-compounds within polymeric matrices. J. Macromol. Sci.: Polym. Rev. **C43**, 547–579 (2003)
63. M.K. Maurer, I.K. Lednev, S.A. Asher, Photoswitchable spirobenzopyran-based photochemically controlled photonic crystals. Adv. Funct. Mater. **15**, 1401–1406 (2005)
64. G.K. Such, R.A. Evans, T.P. Davis, Rapid photochromic switching in a rigid polymer matrix using living radical polymerization. Macromolecules **39**, 1391–1396 (2006)
65. G.K. Such, R.A. Evans, T.P. Davis, The use of block copolymers to systematically modify photochromic behavior. Macromolecules **39**, 9562–9570 (2006)
66. G. Berkovic, V. Krongauz, V. Weiss, Spiropyrans and spirooxazines for memories and switches. Chem. Rev. **100**, 1741–1753 (2000)
67. S.Z. Janicki, G.B. Schuster, A liquid-crystal optooptical switch—nondestructive information-retrieval based on a photochromic fulgide as trigger. J. Am. Chem. Soc. **117**, 8524–8527 (1995)
68. W. Ji, N. Li, D. Chen, et al., Coumarin-containing photo-responsive nanocomposites for NIR light-triggered controlled drug release via a two-photon process. J. Mater. Chem. B **1**, 5942–5949 (2013)
69. N. Fomina, C.L. Mcfearin, M. Sermsakdi, et al., Low power, biologically benign NIR light triggers polymer disassembly. Macromolecules **44**, 8590–8597 (2011)
70. Q. Jin, X. Liu, G. Liu, et al., Fabrication of core or shell reversibly photo cross-linked micelles and nanogels from double responsive water-soluble block copolymers. Polymer **51**, 1311–1319 (2010)
71. M. Nagata, Y. Yamamoto, Photoreversible poly(ethylene glycol)s with pendent coumarin group and their hydrogels. React. Funct. Polym. **68**, 915–921 (2008)
72. S.R. Trenor, A.R. Shultz, B.J. Love, et al., Coumarins in polymers: from light harvesting to photo-cross-linkable tissue scaffolds. Chem. Rev. **104**, 3059–3077 (2004)
73. B.G. Lake, Coumarin metabolism, toxicity and carcinogenicity: relevance for human risk assessment. Food Chem. Toxicol. **37**, 423–453 (1999)
74. C. Zhu, C.J. Bettinger, Light-induced disintegration of robust physically cross-linked polymer networks. Macromol. Rapid Commun. **34**, 1446–1451 (2013)
75. H. Zhao, E.S. Sterner, E.B. Coughlin, et al., o-Nitrobenzyl alcohol derivatives: opportunities in polymer and materials science. Macromolecules **45**, 1723–1736 (2012)
76. Thomas SW, Iii (2012) New applications of photolabile nitrobenzyl groups in polymers. Macromol. Chem. Phys. 213:2443–2449

77. Gumbley P, Koylu D, Thomas SW, Iii (2011) Photoresponsive polymers containing nitrobenzyl esters via ring-opening metathesis polymerization. Macromolecules 44:7956–7961
78. O. Bertrand, J.-M. Schumers, C. Kuppan, et al., Photo-induced micellization of block copolymers bearing 4,5-dimethoxy-2-nitrobenzyl side groups. Soft Matter **7**, 6891–6896 (2011)
79. I. Aujard, C. Benbrahim, M. Gouget, et al., o-Nitrobenzyl photolabile protecting groups with red-shifted absorption: syntheses and uncaging cross-sections for one- and two-photon excitation. Chem. A Eur. J. **12**, 6865–6879 (2006)
80. S. Wang, M.-S. Choi, S.-H. Kim, Bistable photoswitching in poly(N-isopropylacrylamide) with spironaphthoxazine hydrogel for optical data storage. J. Photochem. Photobiol. A: Chem. **198**, 150–155 (2008)
81. L. Zhu, W. Wu, M.-Q. Zhu, et al., Reversibly photoswitchable dual-color fluorescent nanoparticles as new tools for live-cell imaging. J. Am. Chem. Soc. **129**, 3524 (2007)
82. Y. Yokoyama, Fulgides for memories and switches. Chem. Rev. **100**, 1717–1740 (2000)
83. M. Irie, T. Fukaminato, K. Matsuda, et al., Photochromism of diarylethene molecules and crystals: memories, switches, and actuators. Chem. Rev. **114**, 12174–12277 (2014)
84. F.M. Andreopoulos, C.R. Deible, M.T. Stauffer, et al., Photoscissable hydrogel synthesis via rapid photopolymerization of novel PEG-based polymers in the absence of photoinitiators. J. Am. Chem. Soc. **118**, 6235–6240 (1996)
85. Y.J. Zheng, F.M. Andreopoulos, M. Micic, et al., A novel photoscissile poly(ethylene glycol)-based hydrogel. Adv. Funct. Mater. **11**, 37–40 (2001)
86. Y.J. Zheng, M. Mieie, S.V. Mello, et al., PEG-based hydrogel synthesis via the photodimerization of anthracene groups. Macromolecules **35**, 5228–5234 (2002)
87. Q. Jin, G. Liu, J. Li, Preparation of reversibly photo-cross-linked nanogels from pH-responsive block copolymers and use as nanoreactors for the synthesis of gold nanoparticles. Eur. Polym. J. **46**, 2120–2128 (2010)
88. D.Y. Wu, S. Meure, D. Solomon, Self-healing polymeric materials: a review of recent developments. Prog. Polym. Sci. **33**, 479–522 (2008)
89. S. Banerjee, R. Tripathy, D. Cozzens, et al., Photoinduced smart, self-healing polymer sealant for photovoltaics. ACS Appl. Mater. Interfaces **7**, 2064–2072 (2015)
90. Snyder EA, Tong TH (2005) Towards novel light-activated shape memory polymer: thermornechanical properties of photo-responsive polymers. In: Ozkan CS, LaVan DA, McNie M, Prasad S editors. Micro- and nanosystems-materials and devices. p 353–358
91. S. Gug, S. Charon, A. Specht, et al., Photolabile glutamate protecting group with high one- and two-photon uncaging efficiencies. ChemBiochem **9**, 1303–1307 (2008)
92. H. Yu, J. Li, D. Wu, et al., Chemistry and biological applications of photo-labile organic molecules. Chem. Soc. Rev. **39**, 464–473 (2010)
93. H. Xu, M.M. Reynolds, K.E. Cook, et al., 2-Hydroxy-5-nitrobenzyl as a Diazeniumdiolate Protecting Group: application in NO-releasing polymers with enhanced biocompatibility. Org. Lett. **10**, 4593–4596 (2008)
94. P. Anilkumar, E. Gravel, I. Theodorou, et al., Nanometric micelles with photo-triggered cytotoxicity. Adv. Funct. Mater. **24**, 5246–5252 (2014)
95. D.Y. Wong, D.R. Griffin, J. Reed, et al., Photodegradable hydrogels to generate positive and negative features over multiple length scales. Macromolecules **43**, 2824–2831 (2010)
96. J.-M. Schumers, O. Bertrand, C.-A. Fustin, et al., Synthesis and self-assembly of diblock copolymers bearing 2-nitrobenzyl photocleavable side groups. J. Polym. Sci. Part A: Polym. Chem. **50**, 599–608 (2012)
97. D. Han, X. Tong, Y. Zhao, Block copolymer micelles with a dual-stimuli-responsive core for fast or slow degradation. Langmuir **28**, 2327–2331 (2012)
98. C.D.G. Lux, C.L. Mcfearin, S. Joshi-Barr, et al., Single UV or near IR triggering event leads to polymer degradation into small molecules. ACS Macro Lett. **1**, 922–926 (2012)

99. B. Yan, J.-C. Boyer, N.R. Branda, et al., Near-infrared light-triggered dissociation of block copolymer micelles using upconverting nanoparticles. J. Am. Chem. Soc. **133**, 19714–19717 (2011)
100. Y.R. Zhao, Q. Zheng, K. Dakin, et al., New caged coumarin fluorophores with extraordinary uncaging cross sections suitable for biological imaging applications. J. Am. Chem. Soc. **126**, 4653–4663 (2004)
101. A.Z. Suzuki, T. Watanabe, M. Kawamoto, et al., Coumarin-4-ylmethoxycarbonyls as phototriggers for alcohols and phenols. Org. Lett. **5**, 4867–4870 (2003)
102. V.R. Shembekar, Y.L. Chen, B.K. Carpenter, et al., A protecting group for carboxylic acids that can be photolyzed by visible light. Biochemistry **44**, 7107–7114 (2005)
103. R.S. Givens, M. Rubina, J. Wirz, Applications of p-hydroxyphenacyl (pHP) and coumarin-4-ylmethyl photoremovable protecting groups. Photochem. Photobiol. Sci. **11**, 472–488 (2012)
104. Y.V. Il'ichev, M.A. Schworer, J. Wirz, Photochemical reaction mechanisms of 2-nitrobenzyl compounds: methyl ethers and caged ATP. J. Am. Chem. Soc. **126**, 4581–4595 (2004)
105. Y. Jiang, P. Wan, H. Xu, et al., Facile reversible UV-controlled and fast transition from emulsion to gel by using a photoresponsive polymer with a Malachite Green Group. Langmuir **25**, 10,134–10,138 (2009)
106. Z. Sekkat, *Photoreactive organic thin films* (Elsevier Science (USA), San Diego, California, 2002)
107. S.K. Yesodha, C.K. Sadashiva Pillai, N. Tsutsumi, Stable polymeric materials for nonlinear optics: a review based on azobenzene systems. Prog. Polym. Sci. **29**, 45–74 (2004)
108. H. Rau, Photoisomerization of azobenzenes, in *Photochemistry and photophysics*, ed. by J. F. Rabek, (CRC Press, Inc., Boca Raton, FL, 1990)
109. A. Goulet-Hanssens, K.L.W. Sun, T.E. Kennedy, et al., Photoreversible surfaces to regulate cell adhesion. Biomacromolecules **13**, 2958–2963 (2012)
110. V.V. Jerca, F.A. Nicolescu, R. Trusca, et al., Oxazoline-functional polymer particles graft with azo-dye. React. Funct. Polym. **71**, 373–379 (2011)
111. L. Brzozowski, E.H. Sargent, Azobenzenes for photonic network applications: third-order nonlinear optical properties. J. Mater. Sci. Mater. Electron. **12**, 483–489 (2001)
112. G.-A. Jaume, V. Dolores, Recent advances towards azobenzene-based light-driven real-time information-transmitting materials. Beilstein J. Org. Chem. **8**, 1003–1017 (2012)
113. V.V. Jerca, F.A. Jerca, I. Rau, et al., Advances in understanding the photoresponsive behavior of azobenzenes substituted with strong electron withdrawing groups. Opt. Mater. **48**, 160–164 (2015)
114. M.M. Lerch, M.J. Hansen, G.M. Van Dam, et al., Emerging targets in photopharmacology. Angew. Chem. Int. Ed. **55**, 10978–10999 (2016)
115. J. Bieth, S.M. Vratsanos, N.H. Wassermann, et al., Photoregulation of biological activity by photochromic reagents. Inactivators of acetylcholinesterase. Biochemistry **12**, 3023–3027 (1973)
116. I. Tochitsky, A. Polosukhina, V.E. Degtyar, et al., Restoring visual function to blind mice with a photoswitch that exploits electrophysiological remodeling of retinal ganglion cells. Neuron **81**, 800–813 (2014)
117. M.A. Kienzler, A. Reiner, E. Trautman, et al., A red-shifted, fast-relaxing azobenzene photoswitch for visible light control of an ionotropic glutamate receptor. J. Am. Chem. Soc. **135**, 17683–17686 (2013)
118. P. Stawski, M. Sumser, D. Trauner, A photochromic agonist of AMPA receptors. Angew. Chem. Int. Ed. **51**, 5748–5751 (2012)
119. R.H. Kramer, D.L. Fortin, D. Trauner, New photochemical tools for controlling neuronal activity. Curr. Opin. Neurobiol. **19**, 544–552 (2009)
120. J.H. Harvey, D. Trauner, Regulating enzymatic activity with a photoswitchable affinity label. ChemBiochem **9**, 191–193 (2008)
121. D.L. Fortin, M.R. Banghart, T.W. Dunn, et al., Photochemical control of endogenous ion channels and cellular excitability. Nat. Methods **5**, 331–338 (2008)

122. J. Broichhagen, N.R. Johnston, Y. Von Ohlen, et al., Allosteric optical control of a class B G-protein-coupled receptor. Angew. Chem. Int. Ed. **55**, 5865–5868 (2016)
123. J. Broichhagen, J.A. Frank, N.R. Johnston, et al., A red-shifted photochromic sulfonylurea for the remote control of pancreatic beta cell function. Chem. Commun. **51**, 6018–6021 (2015)
124. M.J. Hansen, W.A. Velema, G. De Bruin, et al., Proteasome inhibitors with photocontrolled activity. ChemBioChem **15**, 2053–2057 (2014)
125. A.F. Kisselev, M. Groettrup, Subunit specific inhibitors of proteasomes and their potential for immunomodulation. Curr. Opin. Chem. Biol. **23**, 16–22 (2014)
126. M. Borowiak, W. Nahaboo, M. Reynders, et al., Photoswitchable inhibitors of microtubule dynamics optically control mitosis and cell death. Cell **162**, 403–411 (2015)
127. A.J. Engdahl, E.A. Torres, S.E. Lock, et al., Synthesis, characterization, and bioactivity of the photoisomerizable tubulin polymerization inhibitor azo-combretastatin A4. Org. Lett. **17**, 4546–4549 (2015)
128. J.E. Sheldon, M.M. Dcona, C.E. Lyons, et al., Photoswitchable anticancer activity via trans-cis isomerization of a combretastatin A-4 analog. Org. Biomol. Chem. **14**, 40–49 (2016)
129. W.A. Velema, J.P. Van Der Berg, M.J. Hansen, et al., Optical control of antibacterial activity. Nat. Chem. **5**, 924–928 (2013)
130. S. Pittolo, X. Gómez-Santacana, K. Eckelt, et al., An allosteric modulator to control endogenous G protein-coupled receptors with light. Nat. Chem. Biol. **10**, 813–815 (2014)
131. M. Stein, A. Breit, T. Fehrentz, et al., Optical control of TRPV1 channels. Angew. Chem. Int. Ed. **52**, 9845–9848 (2013)
132. H. Nishioka, X. Liang, H. Asanuma, Effect of the ortho modification of azobenzene on the photoregulatory efficiency of DNA hybridization and the thermal stability of its cis form. Chem. A Eur. J. **16**, 2054–2062 (2010)
133. H. Ito, X. Liang, H. Nishioka, et al., Construction of photoresponsive RNA for photoswitching RNA hybridization. Org. Biomol. Chem. **8**, 5519–5524 (2010)
134. H. Asanuma, X. Liang, H. Nishioka, et al., Synthesis of azobenzene-tethered DNA for reversible photo-regulation of DNA functions: hybridization and transcription. Nat. Protoc. **2**, 203–212 (2007)
135. R.P. Sinha, D.P. Hader, UV-induced DNA damage and repair: a review. Photochem. Photobiol. Sci. **1**, 225–236 (2002)
136. H. Asanuma, T. Ito, T. Yoshida, et al., Photoregulation of the formation and dissociation of a DNA duplex by using the cis-trans isomerization of azobenzene. Angew. Chem. Int. Ed. **38**, 2393–2395 (1999)
137. H. Kang, H. Liu, X. Zhang, et al., Photoresponsive DNA-cross-linked hydrogels for controllable release and cancer therapy. Langmuir **27**, 399–408 (2011)
138. D. Han, J. Huang, Z. Zhu, et al., Molecular engineering of photoresponsive three-dimensional DNA nanostructures. Chem. Commun. **47**, 4670–4672 (2011)
139. R.H. Bisby, C. Mead, C.C. Morgan, Wavelength-programmed solute release from photosensitive liposomes. Biochem. Biophys. Res. Commun. **276**, 169–173 (2000)
140. M. Elsabahy, K.L. Wooley, Design of polymeric nanoparticles for biomedical delivery applications. Chem. Soc. Rev. **41**, 2545–2561 (2012)
141. D. Seliktar, Designing cell-compatible hydrogels for biomedical applications. Science **336**, 1124–1128 (2012)
142. A.M. Kloxin, A.M. Kasko, C.N. Salinas, et al., Photodegradable hydrogels for dynamic tuning of physical and chemical properties. Science **324**, 59–63 (2009)
143. K. Peng, I. Tomatsu, A. Kros, Light controlled protein release from a supramolecular hydrogel. Chem. Commun. **46**, 4094–4096 (2010)
144. D.J. Siegwart, J.K. Oh, K. Matyjaszewski, ATRP in the design of functional materials for biomedical applications. Prog. Polym. Sci. **37**, 18–37 (2012)
145. K. Matyjaszewski, J. Spanswick, Controlled/living radical polymerization. Mater. Today **8**, 26–33 (2005)

146. H.S. Bisht, A.K. Chatterjee, Living free-radical polymerization—a review. J. Macromol. Sci. Part C **41**, 139–173 (2001)
147. Y. Zhao, Rational design of light-controllable polymer micelles. Chem. Rec. **7**, 286–294 (2007)
148. O. Boissiere, D. Han, L. Tremblay, et al., Flower micelles of poly(N-isopropylacrylamide) with azobenzene moieties regularly inserted into the main chain. Soft Mater **7**, 9410–9415 (2011)
149. J.-H. Liu, Y.-H. Chiu, Behaviors of self-assembled diblock copolymer with pendant photosensitive azobenzene segments. J. Polym. Sci. Part A: Polym. Chem. **48**, 1142–1148 (2010)
150. Z. Feng, L. Lin, Z. Yan, et al., Dual responsive block copolymer micelles functionalized by NIPAM and azobenzene. Macromol. Rapid Commun. **31**, 640–644 (2010)
151. D. Wang, J. Liu, G. Ye, et al., Amphiphilic block copolymers bearing strong push-pull azo chromophores: synthesis, micelle formation and photoinduced shape deformation. Polymer **50**, 418–427 (2009)
152. X. Tang, L. Gao, X. Fan, et al., Self-assembly and photoresponsivity property of amphiphilic ABA triblock copolymers containing azobenzene moieties in dilute solution. Macromol. Chem. Phys. **210**, 1556–1562 (2009)
153. G. Wang, X. Tong, Y. Zhao, Preparation of azobenzene-containing amphiphilic diblock copolymers for light-responsive micellar aggregates. Macromolecules **37**, 8911–8917 (2004)
154. S. Son, E. Shin, B.-S. Kim, Light-responsive micelles of spiropyran initiated hyperbranched polyglycerol for smart drug delivery. Biomacromolecules **15**, 628–634 (2014)
155. H.-Y. Lee, K.K. Diehn, K. Sun, et al., Reversible photorheological fluids based on spiropyran-doped reverse micelles. J. Am. Chem. Soc. **133**, 8461–8463 (2011)
156. H.-I. Lee, W. Wu, J.K. Oh, et al., Light-induced reversible formation of polymeric micelles. Angew. Chem. Int. Ed. **46**, 2453–2457 (2007)
157. V.K. Kotharangannagari, A. Sanchez-Ferrer, J. Ruokolainen, et al., Photoresponsive reversible aggregation and dissolution of rod-coil polypeptide diblock copolymers. Macromolecules **44**, 4569–4573 (2011)
158. S.-J. Lim, C.-J. Carling, C.C. Warford, et al., Multifunctional photo- and thermo-responsive copolymer nanoparticles. Dyes Pigments **89**, 230–235 (2011)
159. Z. Chen, Y. He, Y. Wang, et al., Amphiphilic diblock copolymer with dithienylethene pendants: synthesis and photo-modulated self-assembly. Macromol. Rapid Commun. **32**, 977–982 (2011)
160. G.-Y. Liu, C.-J. Chen, D.-D. Li, et al., Near-infrared light-sensitive micelles for enhanced intracellular drug delivery. J. Mater. Chem. **22**, 16865–16871 (2012)
161. C.-J. Chen, G.-Y. Liu, Y.-T. Shi, et al., Biocompatible micelles based on comb-like PEG derivates: formation, characterization, and photo-responsiveness. Macromol. Rapid Commun. **32**, 1077–1081 (2011)
162. K. Szczubialka, I. Moczek, S. Blaszkiewicz, et al., Photocrosslinkable smart terpolymers responding to pH, temperature, and ionic strength. J. Polym. Sci. Part A: Polym. Chem. **42**, 3879–3886 (2004)
163. J.F. Ding, G.J. Liu, Polystyrene block poly(2-cinnamoylethyl methacrylate) nanospheres with cross-linked shells. Macromolecules **31**, 6554–6558 (1998)
164. A. Guo, G.J. Liu, J. Tao, Star polymers and nanospheres from cross-linkable diblock copolymers. Macromolecules **29**, 2487–2493 (1996)
165. S.-I. Yusa, M. Sugahara, T. Endo, et al., Preparation and characterization of a pH-responsive nanogel based on a photo-cross-linked micelle formed from block copolymers with controlled structure. Langmuir **25**, 5258–5265 (2009)
166. J. He, X. Tong, Y. Zhao, Photoresponsive nanogels based on photocontrollable cross-links. Macromolecules **42**, 4845–4852 (2009)
167. J. He, L. Tremblay, S. Lacelle, et al., Preparation of polymer single chain nanoparticles using intramolecular photodimerization of coumarin. Soft Matter **7**, 2380–2386 (2011)

168. J. He, B. Yan, L. Tremblay, et al., Both core- and shell-cross-linked nanogels: photoinduced size change, intraparticle LCST, and interparticle UCST thermal behaviors. Langmuir **27**, 436–444 (2011)
169. J. Jiang, B. Qi, M. Lepage, et al., Polymer micelles stabilization on demand through reversible photo-cross-linking. Macromolecules **40**, 790–792 (2007)
170. G. Liu, C.-M. Dong, Photoresponsive poly(S-(o-nitrobenzyl)-L-cysteine)-b-PEO from a L-cysteine N-carboxyanhydride monomer: synthesis, self-assembly, and phototriggered drug release. Biomacromolecules **13**, 1573–1583 (2012)
171. J.Q. Jiang, X. Tong, D. Morris, et al., Toward photocontrolled release using light-dissociable block copolymer micelles. Macromolecules **39**, 4633–4640 (2006)
172. J. Babin, M. Pelletier, M. Lepage, et al., A new two-photon-sensitive block copolymer nanocarrier. Angew. Chem. Int. Ed. **48**, 3329–3332 (2009)
173. C. Gong, K.-L. Wong, M.H.W. Lam, Photoresponsive molecularly imprinted hydrogels for the photoregulated release and uptake of pharmaceuticals in the aqueous media. Chem. Mater. **20**, 1353–1358 (2008)
174. C. Gomy, A.R. Schmitzer, Synthesis and photoresponsive properties of a molecularly imprinted polymer. Org. Lett. **9**, 3865–3868 (2007)
175. N. Minoura, K. Idei, A. Rachkov, et al., Preparation of azobenzene-containing polymer membranes that function in photoregulated molecular recognition. Macromolecules **37**, 9571–9576 (2004)
176. C. Gong, M.H.-W. Lam, H. Yu, The fabrication of a photoresponsive molecularly imprinted polymer for the photoregulated uptake and release of caffeine. Adv. Funct. Mater. **16**, 1759–1767 (2006)
177. Q. Tang, C. Gong, M.H.-W. Lam, et al., Photoregulated uptake and release of drug by an organic-inorganic hybrid sol-gel material. J. Sol-Gel Sci. Technol. **59**, 495–504 (2011)
178. K. Sumaru, K. Ohi, T. Takagi, et al., Photoresponsive properties of poly(N-isopropylacrylamide) hydrogel partly modified with spirobenzopyran. Langmuir **22**, 4353–4356 (2006)
179. E.U. Kulawardana, T. Kuruwita-Mudiyanselage, D.C. Neckers, Dual Responsive poly(N-isopropylacrylamide) hydrogels having spironaphthoxazines as pendant groups. J. Polym. Sci. Part A: Polym. Chem. **47**, 3318–3325 (2009)
180. E.S. Gil, S.M. Hudson, Stimuli-responsive polymers and their bioconjugates. Prog. Polym. Sci. **29**, 1173–1222 (2004)
181. F.M. Andreopoulos, E.J. Beckman, A.J. Russell, Photoswitchable PEG-CA hydrogels and factors that affect their photosensitivity. J. Polym. Sci. Part A: Polym. Chem. **38**, 1466–1476 (2000)
182. F.M. Andreopoulos, E.J. Beckman, A.J. Russell, Light-induced tailoring of PEG-hydrogel properties. Biomaterials **19**, 1343–1352 (1998)
183. D. Chen, H. Liu, T. Kobayashi, et al., Multiresponsive reversible gels based on a carboxylic azo polymer. J. Mater. Chem. **20**, 3610–3614 (2010)
184. J. Liu, J. Nie, Y. Zhao, et al., Preparation and properties of different photoresponsive hydrogels modulated with UV and visible light irradiation. J. Photochem. Photobiol. A Chem. **211**, 20–25 (2010)
185. S. Patnaik, A.K. Sharma, B.S. Garg, et al., Photoregulation of drug release in azo-dextran nanogels. Int. J. Pharm. **342**, 184–193 (2007)
186. B.V.K.J. Schmidt, M. Hetzer, H. Ritter, et al., Complex macromolecular architecture design via cyclodextrin host/guest complexes. Prog. Polym. Sci. **39**, 235–249 (2014)
187. K. Peng, C. Cui, I. Tomatsu, et al., Cyclodextrin/dextran based drug carriers for a controlled release of hydrophobic drugs in zebrafish embryos. Soft Matter **6**, 3778–3783 (2010)
188. W. Xiao, W.-H. Chen, J. Zhang, et al., Design of a photoswitchable hollow microcapsular drug delivery system by using a supramolecular drug-loading approach. J. Phys. Chem. B **115**, 13796–13802 (2011)

189. X. Hu, P.J. Zheng, X.Y. Zhao, et al., Preparation, characterization and novel photoregulated rheological properties of azobenzene functionalized cellulose derivatives and their at-CD complexes. Polymer **45**, 6219–6225 (2004)
190. T. Sakai, H. Murayama, S. Nagano, et al., Photoresponsive slide-ring gel. Adv. Mater. **19**, 2023 (2007)
191. Y. Takashima, T. Nakayama, M. Miyauchi, et al., Complex formation and gelation between copolymers containing pendant azobenzene groups and cyclodextrin polymers. Chem. Lett. **33**, 890–891 (2004)
192. S. Tamesue, Y. Takashima, H. Yamaguchi, et al., Photoswitchable supramolecular hydrogels formed by cyclodextrins and azobenzene polymers. Angew. Chem. Int. Ed. **49**, 7461–7464 (2010)
193. C.-J. Chen, G.-Y. Liu, X.-S. Liu, et al., Construction of photo-responsive micelles from azobenzene-modified hyperbranched polyphosphates and study of their reversible self-assembly and disassembly behaviours. New J. Chem. **36**, 694–701 (2012)
194. Q. Yan, Y. Xin, R. Zhou, et al., Light-controlled smart nanotubes based on the orthogonal assembly of two homopolymers. Chem. Commun. **47**, 9594–9596 (2011)
195. M.A. Hahn, A.K. Singh, P. Sharma, et al., Nanoparticles as contrast agents for in-vivo bioimaging: current status and future perspectives. Anal. Bioanal. Chem. **399**, 3–27 (2011)
196. M.J. Chen, M.Z. Yin, Design and development of fluorescent nanostructures for bioimaging. Prog. Polym. Sci. **39**, 365–395 (2014)
197. W. Liu, W.S. Zhang, X.Q. Yu, et al., Synthesis and biomedical applications of fluorescent nanogels. Polym. Chem. **7**, 5749–5762 (2016)
198. A.A. Beharry, L. Wong, V. Tropepe, et al., Fluorescence imaging of azobenzene photoswitching in vivo. Angew. Chem. **50**, 1325–1327 (2011)
199. A. Chevalier, W. Piao, K. Hanaoka, et al., Azobenzene-caged sulforhodamine dyes: a novel class of 'turn-on' reactive probes for hypoxic tumor cell imaging. Methods Appl. Fluoresc. **3** (2015)
200. S. Mao, R.K.P. Benninger, Y. Yan, et al., Optical lock-in detection of FRET using synthetic and genetically encoded optical switches. Biophys. J. **94**, 4515–4524 (2008)
201. S. Wan, Y. Zheng, J. Shen, et al., "On–off–on" switchable sensor: a fluorescent spiropyran responds to extreme pH conditions and its bioimaging applications. ACS Appl. Mater. Interfaces **6**, 19515–19519 (2014)
202. Q. Hu, Y.Q. Tan, M. Liu, et al., A new highly selective and sensitive fluorescent probe for Zn2+ and its application in cell-imaging. Dyes Pigments **107**, 45–50 (2014)
203. B. Sen, S.K. Sheet, R. Thounaojam, et al., A coumarin based Schiff base probe for selective fluorescence detection of Al3+ and its application in live cell imaging. Spectrochim. Acta A **173**, 537–543 (2017)
204. L. Huang, J. Cheng, K. Xie, et al., Cu2 + −selective fluorescent chemosensor based on coumarin and its application in bioimaging. Dalton Trans. **40**, 10815–10817 (2011)
205. O. Garcia-Beltran, B.K. Cassels, C. Perez, et al., Coumarin-based fluorescent probes for dual recognition of copper(II) and iron(III) ions and their application in bio-imaging. Sensors (Basel) **14**, 1358–1371 (2014)
206. Q. Sun, W.B. Zhang, J.H. Qian, A ratiometric fluorescence probe for selective detection of sulfite and its application in realistic samples. Talanta **162**, 107–113 (2017)
207. Y.T. Yang, B.Z. Bai, M. Jin, et al., Fluorescent imaging of Au3+ in living cells with two new high selective Au3+ probes. Biosens. Bioelectron. **86**, 939–943 (2016)
208. K.M. Xiong, F.J. Huo, C.X. Yin, et al., A off-on green fluorescent chemosensor for cyanide based on a hybrid coumarin-hemicyanine dye and its bioimaging. Sensors Actuat. B: Chem. **220**, 822–828 (2015)
209. Y. Zhou, Y.W. Yao, J.Y. Li, et al., Nitroxyl induced fluorescence enhancement via reduction of a copper(II) coumarin-ester complex: its application for bioimaging in vivo. Sensors Actuat. B: Chem. **174**, 414–420 (2012)

210. Y.C. Liao, P. Venkatesan, L.F. Wei, et al., A coumarin-based fluorescent probe for thiols and its application in cell imaging. Sensors Actuat. B: Chem. **232**, 732–737 (2016)
211. Y.T. Yang, F.J. Huo, C.X. Yin, et al., Thiol-chromene click chemistry: a coumarin-based derivative and its use as regenerable thiol probe and in bioimaging applications. Biosens. Bioelectron. **47**, 300–306 (2013)
212. C.Y. Chen, W. Liu, C. Xu, et al., A colorimetric and fluorescent probe for detecting intracellular GSH. Biosens. Bioelectron. **71**, 68–74 (2015)
213. W.M. Liu, B.J. Zhou, G.L. Niu, et al., Deep-red emissive crescent-shaped fluorescent dyes: substituent effect on live cell imaging. ACS Appl. Mater. Interfaces **7**, 7421–7427 (2015)
214. T.J. Chozinski, L.A. Gagnon, J.C. Vaughan, Twinkle, twinkle little star: Photoswitchable fluorophores for super-resolution imaging. FEBS Lett. **588**, 3603–3612 (2014)
215. H. Schill, S. Nizamov, F. Bottanelli, et al., 4-Trifluoromethyl-substituted coumarins with large stokes shifts: synthesis, bioconjugates, and their use in super-resolution fluorescence microscopy. Chem. A Eur. J. **19**, 16556–16565 (2013)
216. H. Zhang, C. Wang, T. Jiang, et al., Microtubule-targetable fluorescent probe: site-specific detection and super-resolution imaging of ultratrace tubulin in microtubules of living cancer cells. Anal. Chem. **87**, 5216–5222 (2015)
217. E. Deniz, M. Tomasulo, J. Cusido, et al., Photoactivatable fluorophores for super-resolution imaging based on oxazine auxochromes. J. Phys. Chem. C **116**, 6058–6068 (2012)
218. J. Qiao, L. Qi, Y. Shen, et al., Thermal responsive fluorescent block copolymer for intracellular temperature sensing. J. Mater. Chem. **22**, 11543–11549 (2012)
219. Y. Wang, C.-Y. Hong, C.-Y. Pan, Spiropyran-based hyperbranched star copolymer: synthesis, phototropy, FRET, and bioapplication. Biomacromolecules **13**, 2585–2593 (2012)
220. V.V. Jerca, F.A. Nicolescu, D.S. Vasilescu, et al., Dispersion polymerization of an azo-monomer and methylmethacrylate in the presence of oxazoline macromonomer. P Soc Photo-Opt Ins **7838** (2010)
221. V. Marturano, P. Cerruti, C. Carfagna, et al., Photo-responsive polymer nanocapsules. Polymer **70**, 222–230 (2015)
222. J. Keyvan Rad, A.R. Mahdavian, H. Salehi-Mobarakeh, et al., FRET phenomenon in photoreversible dual-color fluorescent polymeric nanoparticles based on azocarbazole/spiropyran derivatives. Macromolecules **49**, 141–152 (2016)
223. M.W. Urban, Stimuli-responsive colloids: from stratified to self-repairing polymeric films and beyond. Curr. Opin. Colloid Interface Sci. **19**, 66–75 (2014)
224. D. Hu, Z. Tian, W. Wu, et al., Photoswitchable nanoparticles enable high-resolution cell imaging: PULSAR microscopy. J. Am. Chem. Soc. **130**, 15279–15281 (2008)
225. Z. Tian, W. Wu, W. Wan, et al., Single-chromophore-based photoswitchable nanoparticles enable dual-alternating-color fluorescence for unambiguous live cell imaging. J. Am. Chem. Soc. **131**, 4245–4252 (2009)
226. A.D.Q. Li, C. Zhan, D. Hu, et al., Photoswitchable nanoprobes offer unlimited brightness in frequency-domain imaging. J. Am. Chem. Soc. **133**, 7628–7631 (2011)
227. W. Denk, J. Strickler, W. Webb, Two-photon laser scanning fluorescence microscopy. Science **248**, 73–76 (1990)
228. S.L. Ashworth, R.M. Sandoval, G.A. Tanner, et al., Two-photon microscopy: visualization of kidney dynamics. Kidney Int. **72**, 416–421 (2007)
229. K. Svoboda, R. Yasuda, Principles of two-photon excitation microscopy and its applications to neuroscience. Neuron **50**, 823–839 (2006)
230. Y. Imanishi, K.H. Lodowski, Y. Koutalos, Two-photon microscopy: shedding light on the chemistry of vision. Biochemistry **46**, 9674–9684 (2007)
231. J.A. Scherschel, M. Rubart, Cardiovascular Imaging Using Two-Photon Microscopy. Microsc. Microanal. **14**, 492–506 (2008)
232. S. Yao, K.D. Belfield, Two-photon fluorescent probes for bioimaging. Eur. J. Org. Chem. **2012**, 3199–3217 (2012)

233. H.M. Kim, B.R. Cho, Small-molecule two-photon probes for bioimaging applications. Chem. Rev. **115**, 5014–5055 (2015)
234. Y.I. Park, K.T. Lee, Y.D. Suh, et al., Upconverting nanoparticles: a versatile platform for wide-field two-photon microscopy and multi-modal in vivo imaging. Chem. Soc. Rev. **44**, 1302–1317 (2015)
235. C.S. Lim, B.R. Cho, Two-photon probes for biomedical imaging. Tetrahedron **71**, 8219–8249 (2015)
236. I.M. Schießl, H. Castrop, Deep insights: intravital imaging with two-photon microscopy. Pflügers Arch.: Eur. J. Physiol. **468**, 1505–1516 (2016)
237. F. Wang, E. Bélanger, M.-E. Paquet, et al., Probing pain pathways with light. Neuroscience **338**, 248–271 (2016)
238. Z. Yang, A. Sharma, J. Qi, et al., Super-resolution fluorescent materials: an insight into design and bioimaging applications. Chem. Soc. Rev. **45**, 4651–4667 (2016)
239. B. Huang, M. Bates, X.W. Zhuang, Super-resolution fluorescence microscopy. Annu. Rev. Biochem. **78**, 993–1016 (2009)
240. T.A. Klar, S.W. Hell, Subdiffraction resolution in far-field fluorescence microscopy. Opt. Lett. **24**, 954–956 (1999)
241. S.W. Hell, J. Wichmann, Breaking the diffraction resolution limit by stimulated-emission - stimulated-emission-depletion fluorescence microscopy. Opt. Lett. **19**, 780–782 (1994)
242. Y. Zhang, C.Y. Ang, M. Li, et al., Polymer-coated hollow mesoporous silica nanoparticles for triple-responsive drug delivery. ACS Appl. Mater. Interfaces **7**, 18,179–18,187 (2015)
243. C.M. Buffinton, K.J. Tong, R.A. Blaho, et al., Comparison of mechanical testing methods for biomaterials: pipette aspiration, nanoindentation, and macroscale testing. J. Mech. Behav. Biomed. Mater. **51**, 367–379 (2015)
244. A.M. Rosales, K.M. Mabry, E.M. Nehls, et al., Photoresponsive elastic properties of azobenzene-containing poly(ethylene-glycol)-based hydrogels. Biomacromolecules **16**, 798–806 (2015)
245. E.R. Draper, R. Schweins, R. Akhtar, et al., Reversible photoreduction as a trigger for photoresponsive gels. Chem. Mater. **28**, 6336–6341 (2016)
246. P.J. Nowatzki, C. Franck, S.A. Maskarinec, et al., Mechanically tunable thin films of photosensitive artificial proteins: preparation and characterization by nanoindentation. Macromolecules **41**, 1839–1845 (2008)

Chapter 2
Polymer Processing Through Multiphoton Absorption

Konstantina Terzaki and Maria Farsari

2.1 Introduction

Multiphoton lithography (MPL) is a 3D printing technique which enables the "direct writing" of computer-designed structures with resolution of a few tens of nanometers [1, 2]. It is based on the phenomenon of multiphoton absorption by photopolymers; the focused beam of an ultrafast laser "directly writes" the 3D structure inside the volume of a transparent material, causing it to absorb two or more photons and polymerize locally. By removing the "unwritten" area, one can obtain 3D copies of the computer design.

MPL was first demonstrated in 1997 [3], and it was immediately embraced by the photonics community for the fabrication of 3D photonic crystals and nanophotonic devices [4, 5]. More recently, it has been applied in more diverse fields such as micro-optics [6, 7], microfluidics [8], biomedical implants [9, 10], and in 3D scaffolds for cell cultures and tissue engineering [11–13].

In this chapter, we will discuss the basic principles of MPL, the equipment used, and the most commonly employed materials. Finally, we will discuss its application in the field of tissue engineering.

K. Terzaki · M. Farsari (✉)
IESL-FORTH, Heraklion, Greece
e-mail: mfarsari@iesl.forth.gr

J. Van Hoorick et al. (eds.), *Polymer and Photonic Materials Towards Biomedical Breakthroughs*, Micro- and Opto-Electronic Materials, Structures, and Systems, https://doi.org/10.1007/978-3-319-75801-5_2

2.2 Basic Principles of MPL

2.2.1 *Multiphoton Absorption*

MPL is based on the phenomenon of multiphoton absorption (MPA). For simplicity reasons, in this tutorial we will concentrate on two-photon absorption (TPA). This is defined as the *simultaneous* absorption of two photons, exciting a molecule from one state (usually the ground state) to a higher energy electronic state [14].The energy difference between these states equals to the sum of the energies of the photons (Fig. 2.1). TPA is a third-order process, it therefore depends on the square of the light intensity, and it is several orders of magnitude weaker than linear absorption. This is the reason that, even though it was theoretically predicted in 1931 by Maria Göppert-Mayer in her doctoral dissertation [15], it was not verified experimentally until 30 years later by Werner Kaiser, when the invention of the laser provided the required high intensities and permitted the generation of two-photon excited fluorescence in a europium-doped crystal [16].

The TPA cross section is measured in GM (Göppert-Mayer), where.

$$1\,\mathrm{GM} = 1 \times 10^{-50}\,\mathrm{cm}^4\,\mathrm{s\,molecules}^{-1}\,\mathrm{photon}^{-1}.$$

There are several techniques to measure the TPA) of a molecule, and the most commonly used are two-photon excited fluorescence and nonlinear transmission (z-scan) [17, 18].

2.2.2 *Equipment and Experimental Setup*

The MPL experimental setup resembles that of a scanning laser microscope, where the laser is used for polymerization rather than vision. A typical experimental setup is shown in Fig. 2.2. The necessary components of a setup are (1) an ultrafast laser, (2) a beam/sample motion system, (3) a microscope objective, (4) a beam intensity

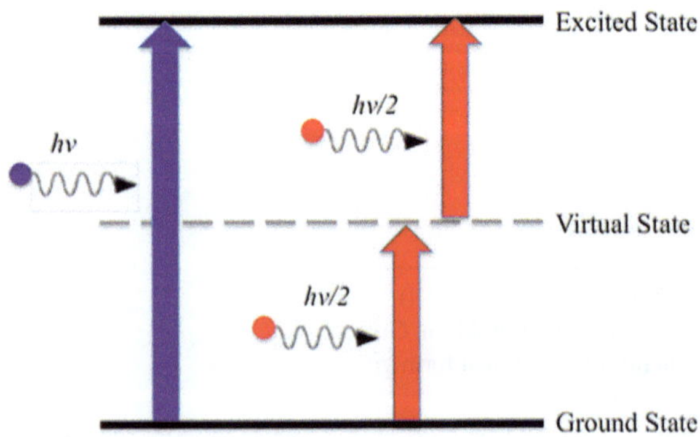

Fig. 2.1 Energy transitions in two-photon absorption

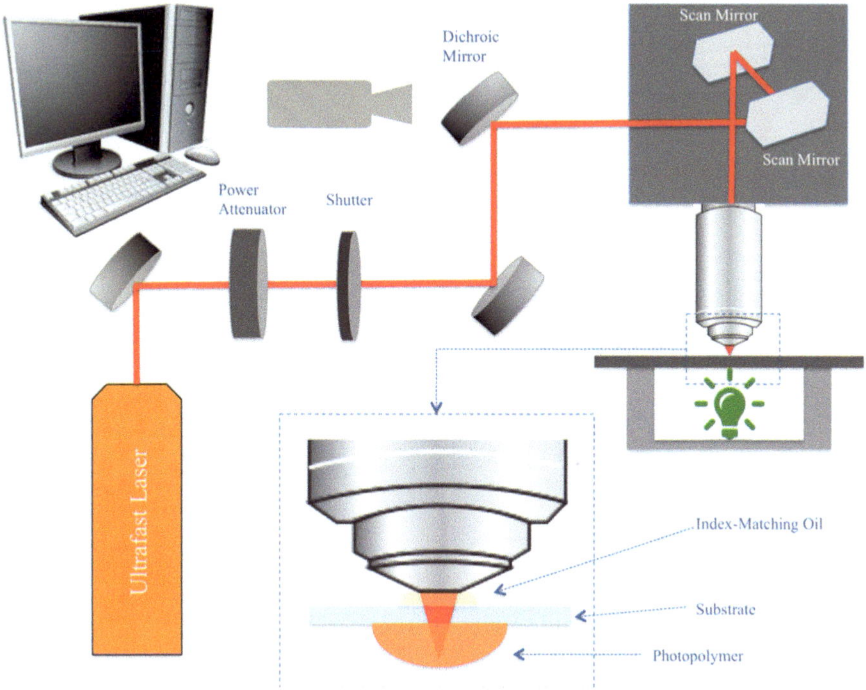

Fig. 2.2 A typical experimental setup consisting of a light source, beam and sample movement components, online camera, beam control, and focusing

control, (5) an online camera, and (6) a control software. These components are detailed below.

1. *Laser*: There are typically three kinds of lasers used: (a) Ti:sapphire femtosecond oscillator (occasionally amplifiers also) operating at around 800 nm; (b) second harmonic fiber lasers, typically operating at 780 nm; and (c) amplified ultrafast YAG (yttrium aluminum garnet) laser operating at the second harmonic (green). As the laser sources vary so much, the energy required for MPL can also vary and depends on the photopolymer, the photoinitiator, and the focusing but also on the laser wavelength and repetition rate.
2. *Motion control*: The "writing" of the structures inside the photopolymer can be done by two methods: (a) The focused laser beam moves inside the photopolymer, using galvanometric scanners. The structure is built in a layer-by-layer format, and after each layer is complete, the sample moves on the z-axis using a linear translation stage. The advantage of this option is speed, as galvanometric scanners can reach speeds of 5 m/s. (b) The beam remains immobile, and the sample moves by high-resolution xyz stages. In this case the stages move in all three directions. The advantage of this method is accuracy, which, depending on the stages used, can be down to nm. In practice, it is very common to have both

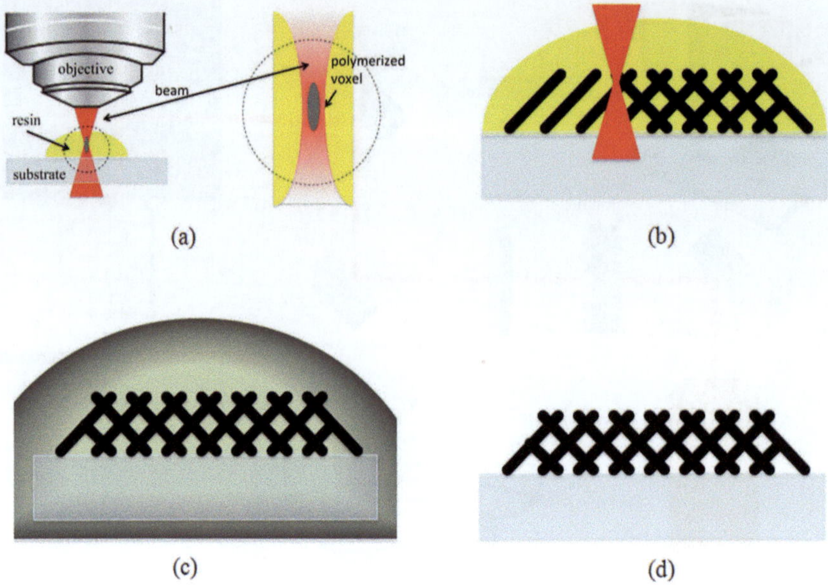

Fig. 2.3 MPL experimental procedure. (**a**) The laser beam is focused inside the volume of the resin. (**b**) The beam moves relatively to the sample following a computer-generated design. (**c**) After laser writing, the sample is immersed into a solvent. (**d**) The freestanding structure is revealed

galvanometric scanners and high-resolution stages in one system and switch between them depending on the requirements of a specific application.

3. *Microscope objective*: The diameter of the focused laser beam is given by the equation

$$d \cong \frac{2\lambda}{\pi \times N.A.}$$

where λ is the laser wavelength and *N.A.* is the numerical aperture of the microscope objective used to focus the laser beam. In order to have high resolution, high *N.A.* objectives need to be employed. If the *N.A.* of the objective is higher than 1, then a matching-index oil is required between the objective and the sample.

4. *Beam intensity control*: The output power of the lasers used in MPL often cannot be controlled directly; it can be done using manual or motorized attenuators (neutral density filters or a combination of a polarizer and a waveplate).
5. *Control software*: The optical and mechanical components described above need to be controlled centrally and synchronized.

It should be noted that nowadays there are several companies providing DLW machines, using various laser sources and scanning configurations [19–22].

The experimental procedure for fabricating a 3D structure by MPL is shown in Fig. 2.3.

2.3 Materials for Laser Polymerization

2.3.1 Introduction

In order for a photopolymer to be suitable for a MPL, it should fulfill the following requirements:

1. It should contain monomers, oligomers, or a mixture of those, which will link through the MPL process to provide the final polymer backbone making up the structure.
2. It should be completely transparent to the wavelength of the laser used to carry out MPL.
3. There should be at least one solvent that dissolves the monomer but not the final polymer, in order to allow structure development and the removal of the unpolymerized resin.
4. The multiphoton polymerization threshold of the resin should be lower than its ablation threshold, for the laser used for MPL.

An MPL resin can be a liquid, a gel, or a solid and can consist of several components, the most common being the monomer/oligomer mixture and a photoinitiator, which will absorb the laser light and provide the active species that will cause this polymerization. Several monomer/oligomer and photoinitiator combinations have been tested. There are also examples of no photoinitiators being used, instead employing redox and Diels-Alder reactions [23, 24].

The materials conventionally used are negative and positive photoresists. When negative photoresists get exposed to multiphoton or UV light, there is a direct writing of the structures in the sample. The polymer chains get cross-linked, making the exposed area insoluble to the solvent used to remove the non-polymerized material. The opposite result occurs in the case of positive photoresists, where multiphoton absorption causes the polymeric chains to break, and thus the irradiated regions get dissolved to the solvent. The negative tone photoresists are subdivided to epoxy- and acrylate-based photoresists [25, 26]. The most well-known commercial materials are the epoxy-based SU-8 [5], Ormocer®, hydrogels [25], and hybrid materials [27, 28]. The positive photoresists are chosen instead of negative photoresists for the simplicity of making structures by fabricating their replicas. Examples of these materials are the AZ Electronic Materials and S1800® [29]. There are also examples of no photoinitiators being used, instead employing redox and Diels-Alder reactions [23, 24]. In the following sections, we will discuss briefly these polymers and photoinitiators.

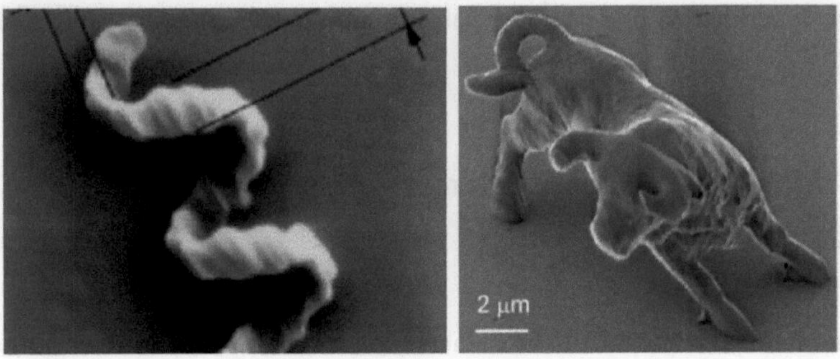

Fig. 2.4 The first 3D structures fabricated by MPL [3, 30]

2.3.2 Acrylate Resins

Acrylate materials are mentioned first for historical reasons; they were the first materials employed for MPL [3, 30] (Fig. 2.4). They are attractive resins, as they are in general low cost and widely available, they dissolve in low-hazard solvents such as alcohols and propanols, they are transparent at visible and near-infrared wavelengths, they exist in a variety of mixtures and formulations, and they can be easily modified to provide functionality. To this end, they have been doped with nanoparticles and quantum dots, [31–33], and with MEH-PPV to make them electroluminescent [34].

2.3.3 SU-8

Among the commercially available materials used in MPL, the solid negative resist SU-8 is unique as it is the only one that employs ions for its cross-linking—all the other widely used materials use radicals. Its maximum absorption wavelength is 365 nm. When exposed to the appropriate light, an acid is generated in the illuminated area; it cross-links and becomes insoluble; this requires heat-treating the samples. As lithography resist, it was designed for the fabrication of high aspect ratio structures.

SU-8 was first structured by MPL by Witzall et al. ([35], Fig. 2.5), and since then, it has been used extensively in MPL fabrication [36–43]. As it is thermally stable, transparent in the visible, and highly resistant to solvents, it is suitable for a variety of applications such as photonic [44], microfluidic [45–47], and biomedical structures [48].

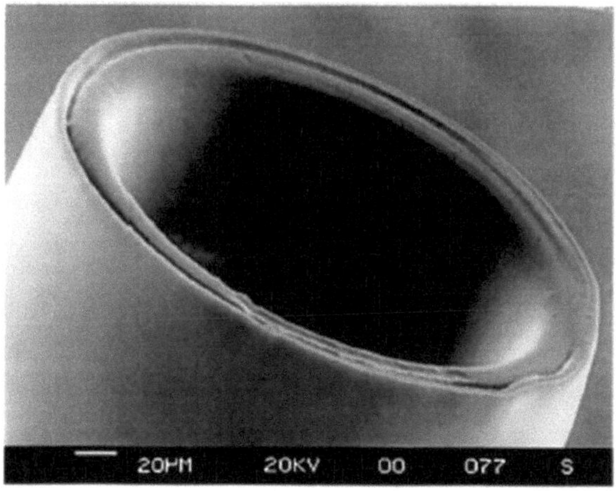

Fig. 2.5 One of the first structures made by MPL using SU-8 as a building material [35]

Fig. 2.6 Micro-optical element made using OrmoComp® [58]

2.3.4 Organic–Inorganic Hybrids

A very popular class of MPL materials are hybrid organic–inorganic photosensitive composites. These materials can be easily prepared via the sol–gel method [27]. They have some very attractive properties, such as chemical, electrochemical, and mechanical stability. In addition, their versatile chemistry allows the incorporation of different materials such as different metal alkoxides, polymers, and even graphene. For these reasons, they have been employed for the fabrication of several microstructures such as photonic components, microfluidic systems, biomedical implants, and cell growth studies (Figs. 2.6 and 2.7).

Silicon-based hybrid composites are very widely used, as they combine all the desirable properties of hybrids, they are optically transparent in visible, and they are also commercially available [49–52]. The most widely used and studied silicate material is the photopolymer OrmoComp® (and its different formulations like Ormocer®, OrmoClear®, etc.), which belongs to the family of organically modified ceramics and is developed by the Fraunhofer Institute for Silicate Research (FhG ISC) and commercially supplied by Microresist Technologies, Germany [53]. It is

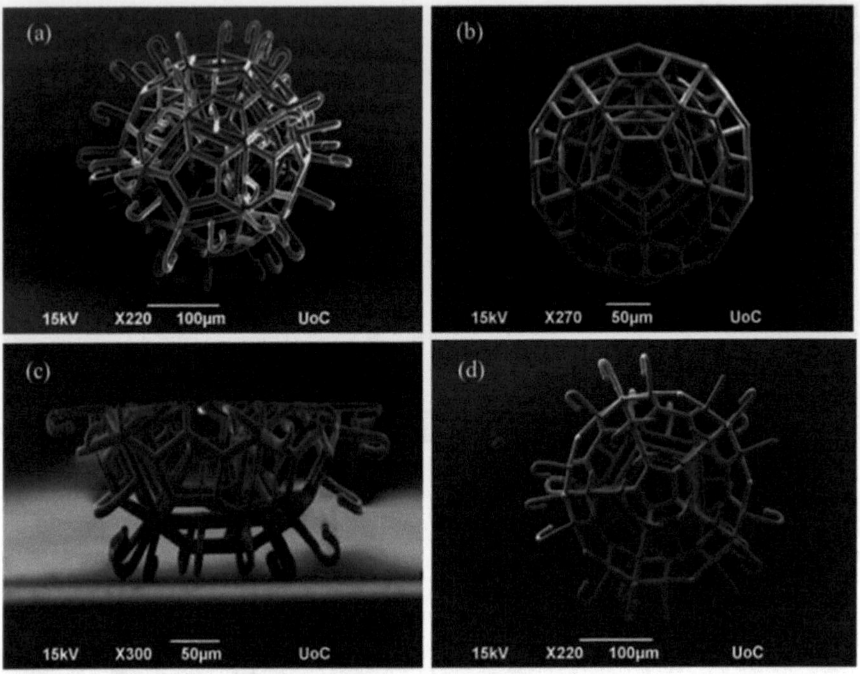

Fig. 2.7 Concentric microscaffolds made using a variant of SZ2080 [81]

available with different material properties as recommended by the application. It has been used for a variety of mostly photonic application but also due to its biocompatibility and chemical inertness as a scaffold for biomolecule immobilization and cell growth applications [48, 51, 54–58], as well as a dental material [59] (http://www.voco.com). Ormocer is a handling, straightforward, and easy material for photopolymerization, but it has a certain drawback according to studies on shrinkage strain evolution during the UV-activated photopolymerization, which showed that as the liquid transforms into a solid during photopolymerization, shrinkage occurs [60].

Combining silicon alkoxides with monomers and other metal alkoxides allows the tailoring of the material properties for specific MPL applications. The most widely known example is the silicon-zirconium hybrid SZ2080, available through IESL-FORTH, Greece [61]. This is a transparent, mechanically stable, biocompatible hybrid that can be structured into complex 3D structures without shrinkage with sub-100-nm resolution [10, 62–74]. Different variants of zirconium-silicon composites have been investigated also, with the addition of the organic monomer 2-(dimethylamino)ethyl methacrylate (DMAEMA) in the initial hybrid material [75–87]. A study on the nanomechanical properties of the materials containing 50% DMAEMA showed excellent mechanical properties, especially after submersion into the cell culture medium [76].

Besides zirconium, other materials have also been employed such as titanium [88], germanium [89], ionogel and grapheme [90], and vanadium [23]. The vanadium case is particularly interesting, as it is the only example to date of redox multiphoton polymerization. There, a vanadium metallo-organic complex initiates the polymerization. It employs multiphoton absorption to self-generate radicals by photoinduced reduction of the metal species from vanadium (V) to vanadium (IV).

2.3.5 *Hydrogels*

MPL-structured hydrogels are used to study cell growth and for tissue engineering applications, as they can mimic the extracellular matrix (ECM) 3D microenvironment [13, 91]. They are roughly divided into three classes: natural materials, modified natural materials, and synthetic hydrogels.

2.3.5.1 Natural Materials

The multiphoton structuring of natural polymer proteins was first demonstrated by Campagnola et al., who worked with a variety of cross-linked proteins such as bovine serum albumin (BSA), fibrinogen and fibronectin, and collagen [92–98]. Unfortunately, due to the poor mechanical properties of these materials, the resulting structures were no more than 2D reliefs. To improve, Seidlits et al. cross-linked protein inside a hyaluronic acid and using this scaffolds for neural cell guidance [99]. Later, they used a picosecond pulse duration green laser to structure pure BSA [100]. BSA and also avidin were cross-linked by Turunen et al. [101] using flavin mononucleotide as a photoinitiator, showing the retention of ligand-binding ability of the cross-linked protein structures. Torgesen et al. photopolymerized BSA encapsulating a live worm [12]. More recently, there have been other examples of MPL structuring of natural materials such as collagen [102] and silk [103].

2.3.5.2 Modified Natural Materials

While natural polymers provide the closest imitation to the ECM, they don't have the mechanical properties required to act as 3D scaffolds. To overcome this problem, researchers chose to modify natural polymers for use in MPL. This is done by adding acrylate and methacrylate groups to natural polymers such as poly-ε-caprolactone [63], gelatin [104, 105] (Fig. 2.8), polylactide [106, 107] (Fig. 2.9), hyaluronic acid [108, 109], and chitosan [109, 110]. This avenue of research has provided the most promising results, as far as biodegradable materials are concerned.

In an earlier study, excellent integrity structures were achieved by doping of acrylic resin with chitosan which did not affect the mechanical properties of the host resin (80 Shore D scale in both cases) [111].

Fig. 2.8 (Left) SEM image of the firstly demonstrated multiphoton structuring of BSA protein. Two BSA stepped pyramids fabricated from a solution of BSA and Rose Bengal as a photosensitizer, with a Ti:sapphire laser [92], (right) 3D woodpile structure fabrication by cross-linking BSA with flavin mononucleotide with Nd:YAG laser

Fig. 2.9 A structure made from methacrylamide-modified gelatin [104]

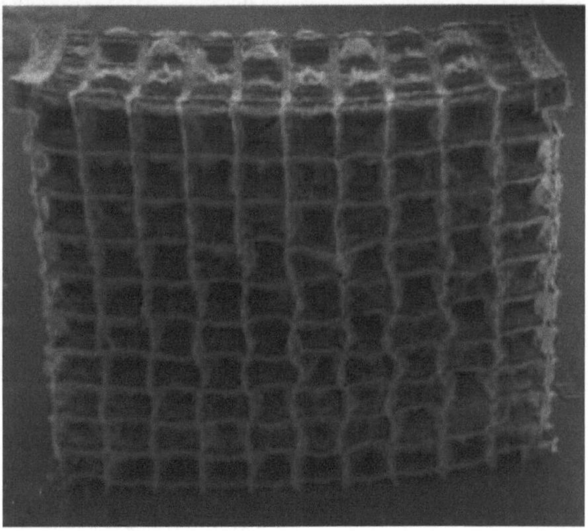

2.3.5.3 Synthetic Hydrogels

The most widely used synthetic hydrogel in tissue engineering applications is PEG (polyethylene glycol) [91]. It is low cost and widely available, is FDA (Food and Drug Administration)-approved, and exists in different formulations. Therefore, cytotoxicity studies on acrylated PEG showed that possible residual materials (e.g., photoinitiator or monomer molecules) that are not consumed during the photopolymerization reaction are toxic to fibroblasts cells, and aging of the scaffolds in aqueous medium is required in order to reduce the cytotoxicity [112]. This could be the reason that its use in MPL has not been very extensive [112–116].

2.3.6 *Photoinitiators*

Until recently, the photoinitiators (PIs) employed in MPL were "traditional" UV photopolymerization PIs. Only recently research started on the synthesis of fast and efficient photoinitiators specifically for multiphoton applications [26, 93, 117–125].

Compared to UV PIs, MPL PIs fulfill also the following requirements [25, 126]:

- Both the PI and the monomer/oligomer are transparent at the wavelength the MPL laser operates.
- The monomer/oligomer is transparent at the two-photon absorption wavelength ($\lambda/2$), to avoid thermal heating and ablation.
- The PI absorbs at the two-photon wavelength and has high two-photon cross section, high radical quantum yield, and highly active radical species generated; typically, if any two of these three are large enough, the initiator will normally be efficient for MPL.

An effective PI is thermally stable and highly soluble in the monomer/oligomer [118]. The most commonly used free radical photoinitiator is benzophenone and its derivatives [63, 127].

2.4 Application in Scaffolds for Cell Growth Studies and Tissue Engineering

In a pioneering article in the journal *Science* in 1993, [128] Langer and Vicanti defined the term "tissue engineering" (TE) as "an interdisciplinary field that applies the principles of engineering and the life sciences toward the development of biological substitutes that restore, maintain, or improve tissue function." They suggest three strategies for TE, the third of them being "Cells placed on or within matrices…cells attached to matrices are implanted and become incorporated into the body. The matrices are fashioned from natural materials such as collagen or from synthetic polymers. Immunological rejection may be prevented by immunosuppressive drugs or by using autologous cells."

Creating a matrix suitable for cell culture and subsequent implantation requires imitating the extracellular matrix using the right biomaterial, cell culture, and growth factors. Micro- and nano-topography have been found to influence cell behavior, like morphology, proliferation, migration, differentiation, and gene expression [129–134], making MPL the ideal technology for the controlled fabrication of the matrix or scaffold, as it is most commonly referred to. For this reason, research into employing MPL as a scaffold fabrication technique has exploded over the last few years (Figs. 2.10, 2.11, 2.12 and 2.13).

To date, the vast majority of studies investigating the use of MPL as a scaffold fabrication technique involved the in vitro study of cell growth, and usually involved

Fig. 2.10 A structure made from photosensitive polylactide [106]

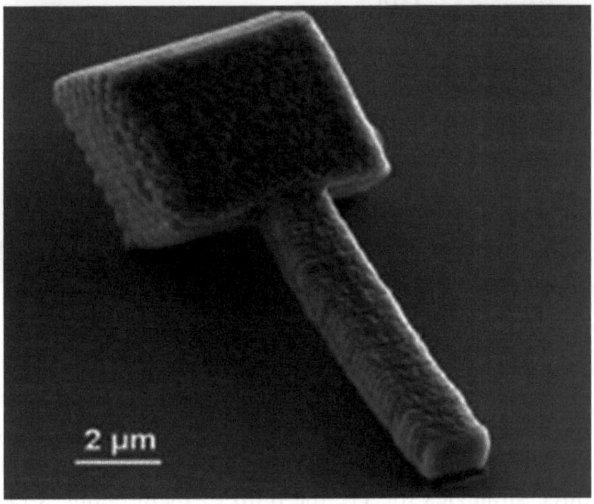

Fig. 2.11 A cantilever made of PEG [114]

the investigation of a specific cell type [76, 82, 83, 106] with a specific photosensitive biomaterial [63, 64, 81, 87, 94, 107, 108, 110, 112] using fairly simple geometries, and occasionally including the use of growth factors [77, 82].

Zirconium-silicon composites have been investigated over the biocompatibility that they exhibit on MC3T3 pre-osteoblastic cells, showing good cell adhesion on the fabricated 3D scaffolds with spread cell morphology, together with a proliferation, reflecting the biocompatibility of the material [76]. Additional studies on the same hybrid photo-structurable material showed that mineralized peptides, when immobilized on the 3D scaffold, strongly support cell adhesion, exhibiting a statistically significant increase of biomineralization, without the addition of growth factors.

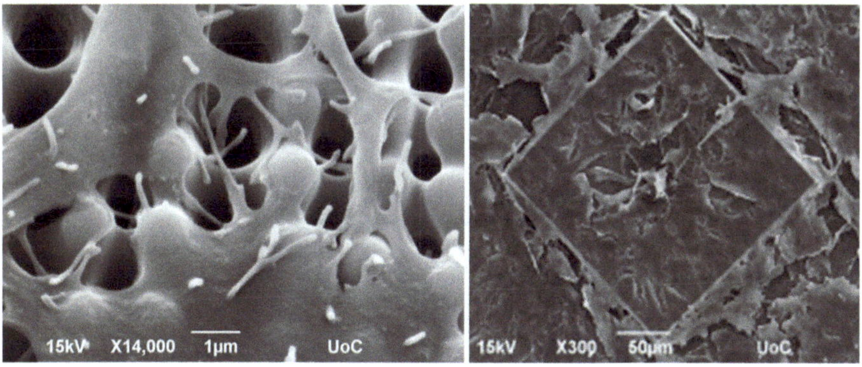

Fig. 2.12 Scanning electron microscopy images of pre-osteoblastic cells cultured on scaffolds of hybrid organic–inorganic material containing DMAEMA. Cells adhere on the scaffolds after 2 h and extend protrusions in the material structures (left); after 72 h the cells proliferated and formed a dense cell layer that completely covered the structures (right) [76]

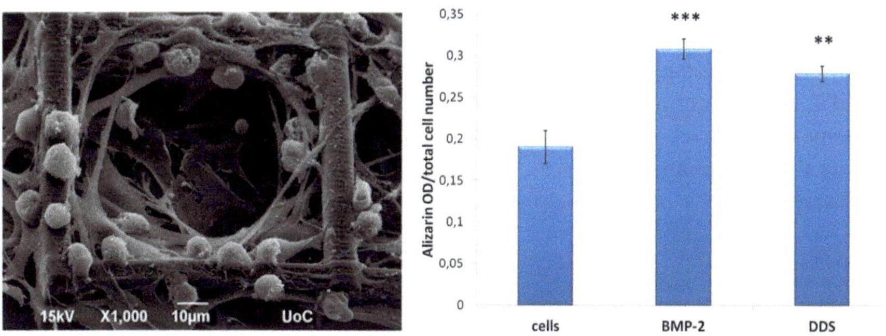

Fig. 2.13 SEM images showing strong pre-osteoblastic cell adhesion cultured on peptide-functionalized scaffolds, after 2 days in culture (left), and quantification of calcium in the ECM of cells normalized to cell number, after 14 days (right) [83]

To the best of our knowledge, the only in vivo implantation of a scaffold made by MPL is described in [68]. The researchers here investigated the suitability of 3D scaffolds made of the hybrid material SZ2080 for cartilage regeneration (Fig. 2.14).

2.5 Conclusions

Polymer processing through multiphoton lithography enables high-quality three-dimensional microstructuring of photosensitive materials using ultrafast laser systems, resulting in high aspect ratio structures. The most widely used commercial materials to tailor-made hybrids used in MPL have been revised, as well as the role of the photoinitiator. The necessity for imitating the extracellular matrix and

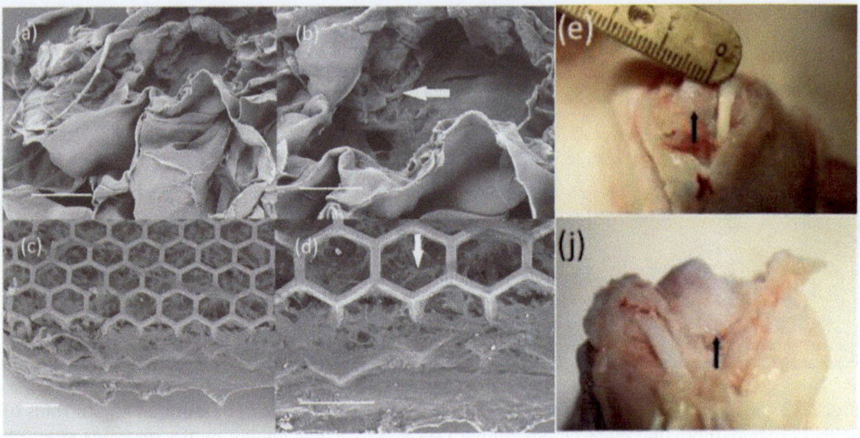

Fig. 2.14 An organic–inorganic hybrid 3D scaffold implanted in rabbit to investigate cartilage regeneration [68]

reproducing the in vivo cellular microenvironment was discussed, emphasizing the use of the photosensitive materials in cell growth studies and tissue engineering.

MPL is a rapidly developing and powerful technology giving the flexibility for implementing novel materials suitable for tissue engineering applications, but despite the encouraging advances, biological studies are scarce and are limited to in vitro cytotoxicity of the developed resins and not to in vivo studies.

References

1. M. Malinauskas, M. Farsari, A. Piskarkas, S. Juodkazis, 3D structuring of transparent materials: a decade of advances. Phys. Rep. 533(1), 1–31 (2013)
2. A. Ovsianikov, V. Mironov, J. Stampfl, R. Liska, Multi-photon processing for applications in biology and tissue engineering. Expert Rev. Med. Dev. **9**, 613–633 (2012)
3. S. Maruo, O. Nakamura, S. Kawata, Three-dimensional microfabrication with two-photon-absorbed photopolymerization. Opt. Lett. **22**, 132–134 (1997)
4. M.M. Hossain, M. Gu, Fabrication methods of 3D periodic metallic nano/microstructures for photonics applications. Las. Photon. Rev. **8**, 233–249 (2014)
5. S. Juodkazis, V. Mizeikis, H. Misawa, Three-dimensional microfabrication of materials by femtosecond lasers for photonics applications. J. Appl. Phys. **106**, 051101 (2009). https://doi.org/10.1063/1.3216462
6. E. Brasselet, M. Malinauskas, A. Zukauskas, S. Juodkazis, Photopolymerized microscopic vortex beam generators: Precise delivery of optical orbital angular momentum. Appl. Phys. Lett. **97**, 211108 (2010). https://doi.org/10.1063/1.3517519
7. M. Malinauskas et al., Femtosecond laser polymerization of hybrid/integrated micro-optical elements and their characterization. J. Opt. **12**, 124010 (2010). https://doi.org/10.1088/2040-8978/12/12/124010
8. L. Amato et al., Integrated three-dimensional filter separates nanoscale from microscale elements in a microfluidic chip. Lab Chip **12**, 1135–1142 (2012). https://doi.org/10.1039/c2lc21116e

9. C. Schizas et al., On the design and fabrication by two-photon polymerization of a readily assembled micro-valve. Int. J. Adv. Manuf. Technol. **48**, 435–441 (2010). https://doi.org/10.1007/s00170-009-2320-4
10. S. Galanopoulos et al., Design, fabrication and computational characterization of a 3D micro-valve built by multi-photon polymerization. Micromachines **5**, 505–514 (2014). https://doi.org/10.3390/mi5030505
11. M.T. Raimondi et al., Two-photon laser polymerization: from fundamentals to biomedical application in tissue engineering and regenerative medicine. J. Appl. Biomater. Fundamental Mater. **10**, 56–66 (2012)
12. J. Torgersen et al., Photo-sensitive hydrogels for three-dimensional laser microfabrication in the presence of whole organisms. J. Biomed. Opt. **17**, 105008 (2012). https://doi.org/10.1117/1.jbo.17.10.105008
13. J. Torgersen et al., Hydrogels for two-photon polymerization: A toolbox for mimicking the extracellular matrix. Adv. Funct. Mater. **23**, 4542–4554 (2013). https://doi.org/10.1002/adfm.201203880
14. R.W. Boyd, *Nonlinear Optics* (Academic Press, Boston, 2008)
15. M. Göppert-Mayer, Über Elementarakte mit zwei Quantensprüngen. Ann. Phys. **401**, 273–294 (1931)
16. W. Kaiser, C.G.B. Garrett, Two-photon excitation in CaF_2:Eu^{2+}. Phys. Rev. Lett. **7**, 229–232 (1961)
17. M. Sheik-Bahae, A.A. Said, T.H. Wei, D.J. Hagan, E.W. Vanstryland, Sensitive measurement of optical nonlinearities using a single beam. IEEE J. Quantum Electron. **26**, 760–769 (1990). https://doi.org/10.1109/3.53394
18. N.S. Makarov, M. Drobizhev, A. Rebane, Two-photon absorption standards in the 550-1600 nm excitation wavelength range. Opt. Express **16**, 4029–4047 (2008). https://doi.org/10.1364/oe.16.004029
19. Nanoscribe. http://www.nanoscribe.de/
20. Workshop of Photonics. http://www.wophotonics.com/
21. Newport. http://www.newport.com/
22. Teem Photonics. http://www.teemphotonics.com/
23. E. Kabouraki et al., Redox multiphoton polymerization for 3D nanofabrication. Nano Lett. **13**, 3831–3835 (2013). https://doi.org/10.1021/nl401853k
24. A.S. Quick et al., Fabrication and spatially resolved functionalization of 3d microstructures via multiphoton-induced diels–Alder chemistry. Adv. Funct. Mater. 24, 3571–3580 (2014). https://doi.org/10.1002/adfm.201304030
25. C.N. LaFratta, J.T. Fourkas, T. Baldacchini, R.A. Farrer, Multiphoton fabrication. Angew. Chem. Int. Ed. **46**, 6238–6258 (2007). https://doi.org/10.1002/anie.200603995
26. M. Farsari, G. Filippidis, K. Sambani, T.S. Drakakis, C. Fotakis, Two-photon polymerization of an eosin Y-sensitized acrylate composite. J. Photochem. Photobio. A: Chem. **181**, 132–135 (2006). https://doi.org/10.1016/j.jphotochem.2005.11.005
27. A. Ovsianikov et al., Shrinkage of microstructures produced by two-photon polymerization of Zr-based hybrid photosensitive materials. Opt. Express **17**, 2143–2148 (2009)
28. M. Farsari, M. Vamvakaki, B.N. Chichkov, Multiphoton polymerization of hybrid materials. J. Opt. **12**, 124001 (2010). https://doi.org/10.1088/2040-8978/12/12/124001
29. D.J.T. Kyle, A. Oikonomou, E. Hill, A. Bayat, Development and functional evaluation of biomimetic silicone surfaces with hierarchical micro/nano-topographical features demonstrates favourable in vitro foreign body response of breast-derived fibroblasts. Biomaterials **52**, 88–102 (2015). https://doi.org/10.1016/j.biomaterials.2015.02.003
30. S. Kawata, H.-B. Sun, T. Tanaka, K. Takada, Finer features for functional microdevices. Nature, 412, 697–698 (2001.) http://www.nature.com/nature/journal/v412/n6848/suppinfo/412697a0_S1.html

31. X.M. Duan, H.B. Sun, K. Kaneko, S. Kawata, Two-photon polymerization of metal ions doped acrylate monomers and oligomers for three-dimensional structure fabrication. Thin Solid Films **453–54**, 518–521 (2004)
32. Z.B. Sun et al., Multicolor polymer nanocomposites: In situ synthesis and fabrication of 3D microstructures. Adv. Mater. **20**, 914–919 (2008). https://doi.org/10.1002/adma.200702035
33. Z.B. Sun et al., Two- and three-dimensional micro/nanostructure patterning of CdS-polymer nanocomposites with a laser interference technique and in situ synthesis. Nanotechnology **19**, 035611 (2008)
34. C.R. Mendonca et al., Three-dimensional fabrication of optically active microstructures containing an electroluminescent polymer. Appl. Phys. Lett. **95**, 113309 (2009). https://doi.org/10.1063/1.3232207
35. G. Witzgall, R. Vrijen, E. Yablonovitch, V. Doan, B.J. Schwartz, Single-shot two-photon exposure of commercial photoresist for the production of three-dimensional structures. Opt. Lett. **23**, 1745–1747 (1998)
36. K.D. Belfield et al., Multiphoton-absorbing organic materials for microfabrication, emerging optical applications and non-destructive three-dimensional imaging. J. Phys. Org. Chem. **13**, 837–849 (2000)
37. S.M. Kuebler et al., Design and application of high-sensitivity two-photon initiators for three-dimensional microfabrication. J. Photochem. Photobio. A: Chem. **158**, 163–170 (2003). https://doi.org/10.1016/s1010-6030(03)00030-3
38. W.H. Teh et al., SU-8 for real three-dimensional subdiffraction-limit two-photon microfabrication. Appl. Phys. Lett. **84**, 4095–4097 (2004). https://doi.org/10.1063/1.1753059
39. V. Mizeikis, K.K. Seet, S. Juodkazis, H. Misawa, Three-dimensional woodpile photonic crystal templates for the infrared spectral range. Opt. Lett. **29**, 2061–2063 (2004)
40. K.K. Seet, V. Mizeikis, S. Juodkazis, H. Misawa, Three-dimensional circular spiral photonic crystal structures recorded by femtosecond pulses. J. Non-Cryst. Solids **352**, 2390–2394 (2006). https://doi.org/10.1016/j.jnoncrysol.2006.02.079
41. B.L. Aekbote et al., Surface-modified complex SU-8 microstructures for indirect optical manipulation of single cells. Biomed. Opt. Express **7**, 45–56 (2016). https://doi.org/10.1364/boe.7.000045
42. W. Horn, S. Kroesen, C. Denz, Two-photon fabrication of organic solid-state distributed feedback lasers in rhodamine 6G doped SU-8. Appl. Phys. B-Lasers Opt. **117**, 311–315 (2014). https://doi.org/10.1007/s00340-014-5837-7
43. M. Licht, A. Uchugonova, K. Konig, M. Straub, Sub-15 fs multiphoton lithography of three-dimensional structures for live cell applications. J. Opt. **14**, 7 (2012). https://doi.org/10.1088/2040-8978/14/6/065601
44. M. Deubel et al., Direct laser writing of three-dimensional photonic-crystal templates for telecommunications. Nat. Mater. **3**, 444–447 (2004). https://doi.org/10.1038/nmat1155
45. D. Wu et al., Femtosecond laser rapid prototyping of nanoshells and suspending components towards microfluidic devices. Lab Chip **9**, 2391–2394 (2009). https://doi.org/10.1039/b902159k
46. G. Kumi, C.O. Yanez, K.D. Belfield, J.T. Fourkas, High-speed multiphoton absorption polymerization: fabrication of microfluidic channels with arbitrary cross-sections and high aspect ratios. Lab Chip **10**, 1057–1060. https://doi.org/10.1039/b923377f
47. M. Stoneman, M. Fox, C.Y. Zeng, V. Raicu, Real-time monitoring of two-photon photopolymerization for use in fabrication of microfluidic devices. Lab Chip **9**, 819–827 (2009). https://doi.org/10.1039/b816993d
48. A. Ovsianikov, S. Schlie, A. Ngezahayo, A. Haverich, B.N. Chichkov, Two-photon polymerization technique for microfabrication of CAD-designed 3D scaffolds from commercially available photosensitive materials. J. Tissue Eng. Regen. Med. **1**, 443–449 (2007). https://doi.org/10.1002/term.57

49. M. Farsari, G. Filippidis, C. Fotakis, Fabrication of three-dimensional structures by three-photon polymerization. Opt. Lett. **30**, 3180–3182 (2005). https://doi.org/10.1364/ol.30.003180
50. Y. Jun, P. Nagpal, D.J. Norris, Thermally stable organic–inorganic hybrid photoresists for fabrication of photonic band gap structures with direct laser writing. Adv. Mater. **20**, 606–610 (2008)
51. V. Dinca et al., Directed three-dimensional patterning of self-assembled peptide fibrils. Nano Lett. **8**, 538–543 (2008). https://doi.org/10.1021/nl072798r
52. A. Matei et al., Functionalized ormosil scaffolds processed by direct laser polymerization for application in tissue engineering. Appl. Surf. Sci. **278**, 357–361 (2013). https://doi.org/10.1016/j.apsusc.2012.10.104
53. http://www.microresist.de/products/ormocers/overview_ormocers_en.htm
54. T.P. Bernat et al., Fabrication of micron-scale cylindrical tubes by two-photon polymerization. Fusion Sci. Technol. **70**, 310–315 (2016). https://doi.org/10.13182/fst15-219
55. M. Bieda, F. Bouchard, A.F. Lasagni, Two-photon polymerization of a branched hollow fiber structure with predefined circular pores. J. Photochem. Photobiol. A-Chem. **319**, 1–7 (2016). https://doi.org/10.1016/j.jphotochem.2015.12.012
56. A. Marino et al., Two-photon lithography of 3D Nanocomposite piezoelectric scaffolds for cell stimulation. ACS Appl. Mater. Interfaces **7**, 25574–25579 (2015). https://doi.org/10.1021/acsami.5b08764
57. E. Kapyla et al., Direct laser writing and geometrical analysis of scaffolds with designed pore architecture for three-dimensional cell culturing. J. Micromech. Microeng. **22**, 13 (2012). https://doi.org/10.1088/0960-1317/22/11/115016
58. E. Harnisch et al., Optimization of hybrid polymer materials for 2PP and fabrication of individually designed hybrid microoptical elements thereof. Opt. Mater. Express **5**, 456–461 (2015). https://doi.org/10.1364/ome.5.000456
59. S. Kalra, A. Singh, M. Gupta, V. Chadha, Ormocer: an aesthetic direct restorative material; an in vitro study comparing the marginal sealing ability of organically modified ceramics and a hybrid composite using an ormocer-based bonding agent and a conventional fifth-generation bonding agent. Contemp. Clin. Dentistry **3**, 48–53 (2012). https://doi.org/10.4103/0976-237x.94546
60. D. Karalekas, C. Schizas, Monitoring of solidification induced strains in two resins used in photofabrication. Mater. Des. **30**, 3705–3712 (2009). https://doi.org/10.1016/j.matdes.2009.02.010
61. M. Farsari, B.N. Chichkov, Materials processing: Two-photon fabrication. Nat Photon **3**, 450–452 (2009)
62. A. Ovsianikov et al., Ultra-low shrinkage hybrid photosensitive material for two-photon polymerization microfabrication. ACS Nano **2**, 2257–2262 (2008). https://doi.org/10.1021/nn800451w
63. F. Claeyssens et al., Three-dimensional biodegradable structures fabricated by two-photon polymerization. Langmuir **25**, 3219–3223 (2009). https://doi.org/10.1021/la803803m
64. S. Psycharakis, A. Tosca, V. Melissinaki, A. Giakoumaki, A. Ranella, Tailor-made three-dimensional hybrid scaffolds for cell cultures. Biomed. Mater. **6**(4), 045008 (2011). https://doi.org/10.1088/1748-6041/6/4/045008
65. M.T. Raimondi et al., Three-dimensional structural niches engineered via two-photon laser polymerization promote stem cell homing. Acta Biomater. **9**, 4579–4584 (2013). https://doi.org/10.1016/j.actbio.2012.08.022
66. M. Malinauskas et al., A femtosecond laser-induced two-photon photopolymerization technique for structuring microlenses. J. Opt. **12**, 035204 (2010). https://doi.org/10.1088/2040-8978/12/3/035204
67. L. Jonusauskas et al., Plasmon assisted 3D microstructuring of gold nanoparticle-doped polymers. Nanotechnology **27**, 154001 (2016). https://doi.org/10.1088/0957-4484/27/15/154001

68. J. Maciulaitis et al., Preclinical study of SZ2080 material 3D microstructured scaffolds for cartilage tissue engineering made by femtosecond direct laser writing lithography. Biofabrication **7**, 015015 (2015). https://doi.org/10.1088/1758-5090/7/1/015015
69. S. Rekstyte, T. Jonavicius, M. Malinauskas, Direct laser writing of microstructures on optically opaque and reflective surfaces. Opt. Lasers Eng. **53**, 90–97 (2014). https://doi.org/10.1016/j.optlaseng.2013.08.017
70. S. Rekstyte, A. Zukauskas, V. Purlys, Y. Gordienko, M. Malinauskas, Direct laser writing of 3D polymer micro/nanostructures on metallic surfaces. Appl. Surf. Sci. **270**, 382–387 (2013). https://doi.org/10.1016/j.apsusc.2013.01.034
71. A. Zukauskas et al., Effect of the photoinitiator presence and exposure conditions on laser-induced damage threshold of ORMOSIL (SZ2080). Opt. Mater. **39**, 224–231 (2015). https://doi.org/10.1016/j.optmat.2014.11.031
72. C.C. Zhang et al., Optimized holographic femtosecond laser patterning method towards rapid integration of high-quality functional devices in microchannels. Sci. Rep. **6**, 9 (2016). https://doi.org/10.1038/srep33281
73. L. Jonusauskas, S. Rekstyte, M. Malinauskas, Augmentation of direct laser writing fabrication throughput for three-dimensional structures by varying focusing conditions. Opt. Eng. **53** (2014). https://doi.org/10.1117/1.oe.53.12.125102
74. M. Manousidaki, D.G. Papazoglou, M. Farsari, S. Tzortzakis, Abruptly autofocusing beams enable advanced multiscale photo-polymerization. Optica **3**, 525–530 (2016). https://doi.org/10.1364/optica.3.000525
75. A. Skarmoutsou et al., Nanomechanical properties of hybrid coatings for bone tissue engineering. J. Mech. Behav. Biomed. Mater. **25**, 48–62 (2013). https://doi.org/10.1016/j.jmbbm.2013.05.003
76. K. Terzaki et al., Pre-osteoblastic cell response on three-dimensional, organic–inorganic hybrid material scaffolds for bone tissue engineering. J. Biomed. Mater. Res. A **101A**, 2283–2294 (2013). https://doi.org/10.1002/jbm.a.34516
77. M. Chatzinikolaidou et al., Recombinant human bone morphogenetic protein 2 (rhBMP-2) immobilized on laser-fabricated 3D scaffolds enhance osteogenesis. Colloids Surf. B: Biointerfaces **149**, 233–242 (2017). https://doi.org/10.1016/j.colsurfb.2016.10.027
78. A.I. Aristov et al., 3D plasmonic crystal metamaterials for ultra-sensitive biosensing. Sci Rep **6**, 25380 (2016)
79. V. Melissinaki, M. Farsari, S. Pissadakis, A fiber-endface, Fabry-Perot vapor microsensor fabricated by multiphoton polymerization. IEEE J. Sel. Top. Quantum Electron. **21**, 5600110 (2015). https://doi.org/10.1109/jstqe.2014.2381463
80. G. Kenanakis et al., A three-dimensional infra-red metamaterial with asymmetric transmission. ACS Photonics **2**, 287–294 (2015)
81. P. Danilevicius et al., Burr-like, laser-made 3D microscaffolds for tissue spheroid encagement. Biointerphases **10**, 021011 (2015). https://doi.org/10.1116/1.4922646
82. M. Chatzinikolaidou et al., Adhesion and growth of human bone marrow mesenchymal stem cells on precise-geometry 3D organic–inorganic composite scaffolds for bone repair. Mater. Sci. Eng.: C **48**, 301–309 (2015). https://doi.org/10.1016/j.msec.2014.12.007
83. K. Terzaki et al., Mineralized self-assembled peptides on 3D laser-made scaffolds: a new route toward 'scaffold on scaffold' hard tissue engineering. Biofabrication **5**, 045002 (2013). https://doi.org/10.1088/1758-5082/5/4/045002
84. N. Vasilantonakis et al., Three-dimensional metallic photonic crystals with optical Bandgaps. Adv. Mater. **24**, 1101–1105 (2012). https://doi.org/10.1002/adma.201104778
85. I. Sakellari et al., Diffusion-assisted high-resolution direct femtosecond laser writing. ACS Nano **6**, 2302–2311 (2012). https://doi.org/10.1021/nn204454c
86. K. Terzaki et al., 3D conducting nanostructures fabricated using direct laser writing. Opt. Mater. Express **1**, 586–597 (2011)
87. S.A. Skoog et al., Two-photon polymerization of 3-D zirconium oxide hybrid scaffolds for long-term stem cell growth. Biointerphases **9**, 7 (2014). https://doi.org/10.1116/1.4873688

88. I. Sakellari et al., Two-photon polymerization of titanium-containing sol–gel composites for three-dimensional structure fabrication. Appl. Phys. A **100**, 359–364 (2010). https://doi.org/10.1007/s00339-010-5864-0
89. M. Malinauskas et al., 3D microoptical elements formed in a photostructurable germanium silicate by direct laser writing. Opt. Lasers Eng. **50**, 1785–1788 (2012). https://doi.org/10.1016/j.optlaseng.2012.07.001
90. M. Oubaha et al., Graphene-doped photo-patternable ionogels: tuning of conductivity and mechanical stability of 3D microstructures. J. Mater. Chem. **22**, 10552–10559 (2012). https://doi.org/10.1039/c2jm30512g
91. J.L. Drury, D.J. Mooney, Hydrogels for tissue engineering: scaffold design variables and applications. Biomaterials **24**, 4337–4351 (2003). https://doi.org/10.1016/S0142-9612(03)00340-5
92. J.D. Pitts, P.J. Campagnola, G.A. Epling, S.L. Goodman, Submicron multiphoton free-form fabrication of proteins and polymers: Studies of reaction efficiencies and applications in sustained release. Macromolecules **33**, 1514–1523 (2000). https://doi.org/10.1021/ma9910437
93. J.D. Pitts et al., New photoactivators for multiphoton excited three-dimensional submicron cross-linking of proteins: Bovine serum albumin and type 1 collagen. Photochem. Photobiol. **76**, 135–144 (2002). https://doi.org/10.1562/0031-8655(2002)076<0135:npfmet>2.0.co;2
94. S. Basu, P.J. Campagnola, Properties of crosslinked protein matrices for tissue engineering applications synthesized by multiphoton excitation. J. Biomed. Mater. Res. A **71A**, 359–368 (2004). https://doi.org/10.1002/jbm.a.30175
95. S. Basu, P.J. Campagnola, Enzymatic activity of alkaline phosphatase inside protein and polymer structures fabricated via multiphoton excitation. Biomacromolecules **5**, 572–579 (2004). https://doi.org/10.1021/bm0344194
96. S. Basu et al., Multiphoton excited fabrication of collagen matrixes cross-linked by a modified benzophenone dimer: Bioactivity and enzymatic degradation. Biomacromolecules **6**, 1465–1474 (2005). https://doi.org/10.1016/bm049258y
97. L.P. Cunningham, M.P. Veilleux, P.J. Campagnola, Freeform multiphoton excited microfabrication for biological applications using a rapid prototyping CAD-based approach. Opt. Exp. **14**, 8613–8621 (2006). https://doi.org/10.1364/oe.14.008613
98. P.J. Su et al., Mesenchymal stem cell interactions with 3D ECM modules fabricated via multiphoton excited photochemistry. Biomacromolecules **13**, 2917–2925 (2012). https://doi.org/10.1021/bm300949k
99. S.K. Seidlits, C.E. Schmidt, J.B. Shear, High-resolution patterning of hydrogels in three dimensions using direct-writep for cell guidance. Adv. Funct. Mater. **19**, 3543–3551 (2009)
100. E.T. Ritschdorff, J.B. Shear, Multiphoton lithography using a high-repetition rate microchip laser. Anal. Chem. **82**, 8733–8737 (2010). https://doi.org/10.1021/ac101274u
101. S. Turunen et al., Pico- and femtosecond laser-induced crosslinking of protein microstructures: evaluation of processability and bioactivity. Biofabrication **3**, 045002 (2011)
102. M.A. Skylar-Scott, M.C. Liu, Y.L. Wu, A. Dixit, M.F. Yanik, Guided homing of cells in multi-photon microfabricated bioscaffolds. Adv. Healthc. Mater. **5**, 1233–1243 (2016). https://doi.org/10.1002/adhm.201600082
103. K. Maximova et al., Silk patterns made by direct femtosecond laser writing. Biomicrofluidics **10**, 054101 (2016)
104. A. Ovsianikov et al., Laser fabrication of three-dimensional CAD scaffolds from photosensitive gelatin for applications in tissue engineering. Biomacromolecules **12**, 851–858 (2011). https://doi.org/10.1021/bm1015305
105. A. Ovsianikov et al., Laser Photofabrication of cell-containing hydrogel constructs. Langmuir **30**, 3787–3794 (2014). https://doi.org/10.1021/la402346z
106. V. Melissinaki et al., Direct laser writing of 3D scaffolds for neural tissue engineering applications. Biofabrication **3**, 045005 (2011). https://doi.org/10.1088/1758-5082/3/4/045005

107. P. Danilevicius et al., The effect of porosity on cell ingrowth into accurately defined, laser-made, polylactide-based 3D scaffolds. Appl. Surf. Sci. **336**, 2–10 (2015)
108. O. Kufelt, A. El-Tamer, C. Sehring, S. Schlie-Wolter, B.N. Chichkov, Hyaluronic acid based materials for scaffolding via two-photon polymerization. Biomacromolecules **15**, 650–659 (2014). https://doi.org/10.1021/bm401712q
109. P.S. Timashev et al., Compatibility of cells of the nervous system with structured biodegradable chitosan-based hydrogel matrices. Appl. Biochem. Microbiol. **52**, 508–514 (2016). https://doi.org/10.1134/s0003683816050161
110. O. Kufelt et al., Water-soluble photopolymerizable chitosan hydrogels for biofabrication via two-photon polymerization. Acta Biomater. **18**, 186–195 (2015). https://doi.org/10.1016/j.actbio.2015.02.025
111. D.S. Correa, P. Tayalia, G. Cosendey, D.S. dos Santos Jr., R.F. Aroca, E. Mazur, C.R. Mendonca, Two-photon polymerization for fabricating structures containing the biopolymer chitosan. J. Nanosci. Nanotechnol. **9**, 5845–5849 (2009)
112. A. Ovsianikov et al., Three-dimensional laser micro- and nano-structuring of acrylated poly(ethylene glycol) materials and evaluation of their cytoxicity for tissue engineering applications. Acta Biomater. **7**, 967–974 (2011). https://doi.org/10.1016/j.actbio.2010.10.023
113. M. Malinauskas et al., 3D artificial polymeric scaffolds for stem cell growth fabricated by femtosecond laser. Lith. J. Phys. **50**, 75–82 (2010). https://doi.org/10.3952/lithjphys.50121
114. T. Honegger, T. Elmberg, K. Berton, D. Peyrade, Visible microlaser two-photon polymerization in a microfludic cell: A resist study. Microel. Engin. **88**, 2725–2728 (2011). https://doi.org/10.1016/j.mee.2010.12.094
115. W.D. Zhang, S.C. Chen, Femtosecond laser nanofabrication of hydrogel biomaterial. MRS Bull. **36**, 1028–1033 (2011). https://doi.org/10.1557/mrs.2011.275
116. W.D. Zhang, P. Soman, K. Meggs, X. Qu, S.C. Chen, Tuning the Poisson's ratio of biomaterials for investigating cellular response. Adv. Funct. Mater. **23**, 3226–3232 (2013). https://doi.org/10.1002/adfm.201202666
117. J. Xing et al., A water soluble initiator prepared through host-guest chemical interaction for microfabrication of 3D hydrogels via two-photon polymerization. J. Mater. Chem. B **2**, 4318–4323 (2014). https://doi.org/10.1039/c4tb00414k
118. Z. Li et al., A straightforward synthesis and structure–activity relationship of highly efficient initiators for two-photon polymerization. Macromolecules **46**, 352–361 (2013). https://doi.org/10.1021/ma301770a
119. R. Nazir, P. Danilevicius, D. Gray, M. Farsari, D.T. Gryko, Push-pull acyl-phosphine oxides for two-photon-induced polymerization. Macromolecules **46**, 7239–7244 (2013)
120. R. Nazir et al., π-expanded keto-coumarins as efficient, biocompatible initiators for two-photon induced polymerization. Chem. Mater. **26**(10), 3175–3184 (2014). https://doi.org/10.1021/cm500612w
121. Z.Q. Li et al., Initiation efficiency and cytotoxicity of novel water-soluble two-photon photoinitiators for direct 3D microfabrication of hydrogels. RSC Adv. **3**, 15939–15946 (2013). https://doi.org/10.1039/c3ra42918k
122. R. Nazir et al., π-Expanded α,β-unsaturated ketones: synthesis, optical properties, and two-photon-induced polymerization. ChemPhysChem **16**(3), 682–690 (2015). https://doi.org/10.1002/cphc.201402646
123. R. Nazir et al., Donor–acceptor type Thioxanthones: synthesis, optical properties, and two-photon induced polymerization. Macromolecules **48**, 2466–2472 (2015). https://doi.org/10.1021/acs.macromol.5b00336
124. W.-E. Lu, X.-Z. Dong, W.-Q. Chen, Z.-S. Zhao, X.-M. Duan, Novel photoinitiator with a radical quenching moiety for confining radical diffusion in two-photon induced Photopolymerization. J. Mater. Chem. **21**, 5650–5659 (2011)
125. R. Nazir et al., π-Expanded 1, 3-diketones–synthesis, optical properties and application in two-photon polymerization. J. Mater. Chem. C **4**, 167–177 (2016)

126. Sun, H.-B. & Kawata, S. in *Twophoton Photopolymerization and 3D Lithographic Microfabrication*, ed. by N. Fatkullin. NMR. 3D Analysis. Photopolymerization, vol 170 (Springer, Berlin, 2004), pp. 169–273
127. B. Bhuian, R.J. Winfield, S. O'Brien, G.M. Crean, Investigation of the two-photon polymerisation of a Zr-based inorganic–organic hybrid material system. Appl. Surf. Sci. **252**, 4845–4849 (2006)
128. R. Langer, J. Vacanti, P. Tissue engineering. Science **260**, 920–926 (1993). https://doi.org/10.1126/science.8493529
129. D.E. Discher, P. Janmey, Y.-L. Wang, Tissue cells feel and respond to the stiffness of their substrate. Science **310**, 1139–1143 (2005). https://doi.org/10.1126/science.1116995
130. A.J. Engler, S. Sen, H.L. Sweeney, D.E. Discher, Matrix elasticity directs stem cell lineage specification. Cell **126**, 677–689 (2006). https://doi.org/10.1016/j.cell.2006.06.044
131. M.J. Dalby et al., Fibroblast reaction to island topography: changes in cytoskeleton and morphology with time. Biomaterials **24**, 927–935 (2003). https://doi.org/10.1016/S0142-9612(02)00427-1
132. M.J. Dalby et al., Nanomechanotransduction and interphase nuclear organization influence on genomic control. J. Cell. Biochem. **102**, 1234–1244 (2007). https://doi.org/10.1002/jcb.21354
133. M.J. Dalby et al., Increasing fibroblast response to materials using nanotopography: Morphological and genetic measurements of cell response to 13-nm-high polymer demixed islands. Exp. Cell Res. **276**, 1–9 (2002). https://doi.org/10.1006/excr.2002.5498
134. V. Karageorgiou, D. Kaplan, Porosity of 3D biomaterial scaffolds and osteogenesis. Biomaterials **26**, 5474–5491 (2005). https://doi.org/10.1016/j.biomaterials.2005.02.002

Chapter 3
Two-Photon Polymerization in Tissue Engineering

Anastasia Shpichka, Anastasia Koroleva, Daria Kuznetsova, Vitaliy Burdukovskii, Boris Chichkov, Viktor Bagratashvili, and Peter Timashev

Abbreviations

2PP	Two-photon polymerization
3D	Three-dimensional
ECM	Extracellular matrix

A. Shpichka
Sechenov First Moscow State Medical University, Institute for Regenerative Medicine, Moscow, Russia

A. Koroleva
Laser Zentrum Hannover e.V, Hannover, Germany
e-mail: a.koroleva@lzh.de

D. Kuznetsova
Nizhny Novgorod State Medical Academy, Institute of Biomedical Technologies, Nizhny Novgorod, Russia

V. Burdukovskii
Baikal Institute of Nature Management, Siberian Branch of the Russian Academy of Sciences, Ulan-Ude, Russia

B. Chichkov
Leibniz Universität Hannover, Institute of Quantum Optics, Hannover, Germany
e-mail: chichkov@iqo.uni-hannover.de

V. Bagratashvili
Research Center "Crystallography and Photonics" RAS, Institute of Photonic Technologies, Moscow, Russia

P. Timashev (✉)
Sechenov First Moscow State Medical University, Institute for Regenerative Medicine, Moscow, Russia

Research Center "Crystallography and Photonics" RAS, Institute of Photonic Technologies, Moscow, Russia

J. Van Hoorick et al. (eds.), *Polymer and Photonic Materials Towards Biomedical Breakthroughs*, Micro- and Opto-Electronic Materials, Structures, and Systems, https://doi.org/10.1007/978-3-319-75801-5_3

TE	Tissue engineering
TPA	Two-photon absorption

3.1 Introduction

One of the key elements of therapeutic approaches in tissue engineering is a three-dimensional structure (scaffold or matrix) formed typically from polymeric, ceramic, or hybrid materials using various methods, particularly, two-photon polymerization.

Scaffold should perform the following tasks:

1. Promote cell attachment, proliferation, and differentiation
2. Deliver and retain cells and growth factors
3. Ensure nutrient and oxygen delivery
4. Ensure optimum mechanical and biological environment for tissue regeneration

To achieve these goals, scaffolds should meet the following biological and physical requirements:

1. Biological requirements:
   - Biocompatibility (material cannot be toxic and should promote cell attachment, proliferation, and differentiation)
   - Biodegradability (material should break down over time into nontoxic compounds)
   - Controlled degradation level (adjustable/adaptable level of degradation)
   - Optimal porosity of scaffold microstructure for better tissue ingrowth and vascularization
2. Physical requirements:
   - Optimum strength and rigidity, which coincide with tissue mechanical properties
   - Suitable surface for biomechanical coupling between scaffolds and surrounding tissues
   - Physical-chemical stability, which ensures safe sterilization without changing scaffold properties

Each characteristic of 3D structures is responsible for providing specific biological effect. For example, biocompatibility contributes to the high cell viability and the formation of an adequate response of recipient tissues upon transplantation [1]. Biodegradability is responsible for tissue remodeling rate [2]. Porosity provides cell migration and vessel formation (vascularization) [3–5]. After the attachment of cells to scaffold surfaces, cohesive forces, transferred to cell nuclei by intracellular mechanisms, launch a number of signals. These signals can trigger the synthesis of specific transcription factors [6]. Therefore, mechanical interaction between ECM and cytoskeleton can induce specific stem cell differentiation [7].

Table 3.1 Comparative characteristics of modern approaches of 3D printing methods used to fabricate scaffolds in TE (by Bártolo and Almeida [8], Raimondi et al. [9])

Technique	Resolution (material dependent)	Fabrication speed (material dependent)	Advantages	Disadvantages
Selective laser sintering	150 μm	Up to 10 m/s [10]	Relatively high scaffold strength No need of toxic solvents Multimaterial	Blockage of fine inner pores by powder High operating temperature
Fused deposition modeling	300 μm	30 m/s [11]	No material debris inside scaffold No need of toxic solvents	High operating temperature Limited structure geometry Low-resolution Post-processing
3D inkjet printing	50 μm	Up to10.000 droplets/s [12]	Low operating temperature simple and low cost process	Blockage of fine inner pores by material debris Low mechanical properties
Bioplotting	100 μm	0.15 m/s [13, 14]	Large material variety for hard and soft TE applications	Limited structure geometry Low resolution
Two-photon polymerization	40 nm	5 m/s [15]	Large material variety High structure resolution High fidelity with computer design	Relatively low fabrication speed Limited to fabrication of small structures

In TE, it is particularly interesting to apply modern fabrication techniques for porous scaffold formation. The most frequently used for TE applications are 3D printing, bioplotting, and two-photon polymerization. Table 3.1 describes advantages and disadvantages of different methods for 3D scaffold structure formation.

Although, in comparison to other techniques, 2PP is relatively slow and not yet capable of large structure formation, this technique allows the use of a large material variety for scaffold fabrication with the possibility of controlling accurately their microarchitecture.

Furthermore, the use of 2PP in combination with other microfabrication methods eliminates some drawbacks. For example, combination with micromolding technique [16, 17] can significantly increase the rate of scaffold reproduction with defined structure and make the application of even more different materials possible. This chapter discusses recent studies on 2PP and examples of its application in TE.

3.2 Two-Photon Polymerization

In the late 1990s of the last century, laser applications reached a new level because of the development of technologies, which provide high accuracy of positioning and energy focusing. Interactions of laser radiation with polymers, as well as material adaptation to novel laser technologies, are important tasks of modern science.

Two-photon polymerization belongs to one of the most popular laser microfabrication methods and was rapidly grown in the last decade [18]. This method allows the creation of 3D micro- and nanostructures of any complexity from photosensitive polymers. A tightly focused laser beam with pulse duration from tens to hundreds of femtoseconds is applied to make 3D structures within photosensitive polymer volume using the nonlinear TPA process (Fig. 3.1a).

In contrast to laser stereolithography, where the absorption of a single photon excites a photoinitiator, simultaneous absorption of two or more photons triggers excitation during 2PP. Most materials, which can be polymerized by exposure to single photon ultraviolet irradiation, may undergo similar reactions when two photons are absorbed simultaneously.

The TPA process depends nonlinearly (quadratically) on the laser pulse intensity, which is at its maximum in the laser beam focus. Polymerization occurs in the region of maximum intensity around the laser focus where excitation energy threshold is reached. This region is called "voxel" (Fig. 3.1b). Because of the nonlinear TPA nature, polymerization voxel size may be less than the diffraction limit [18].

Figure 3.2 represents schematic 2PP setup. Femtosecond generator is used as a source of infrared laser pulses with a repetition rate of about 80 MHz and an average laser pulse output power of 300 mW. Both the lower repetition rate and laser power limit the structuring dynamic range and speed. The half-wave plate is used to change the polarization of linearly polarized laser light. In combination with the subsequent

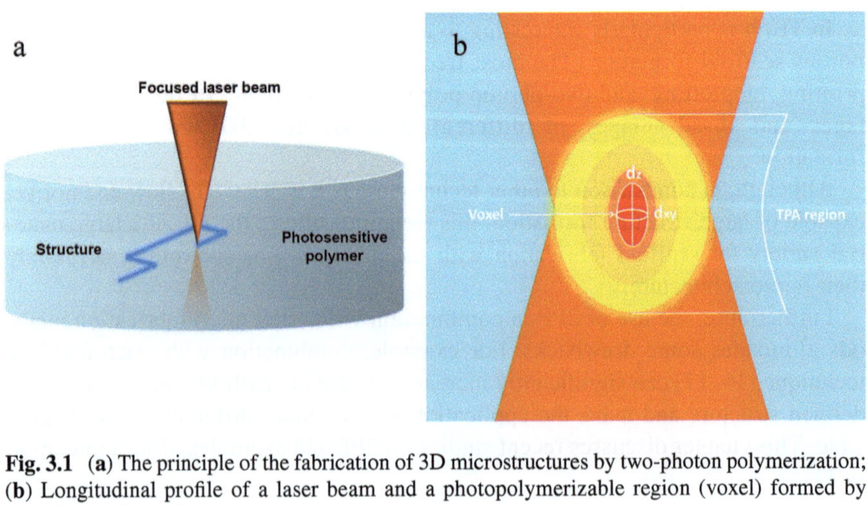

Fig. 3.1 (**a**) The principle of the fabrication of 3D microstructures by two-photon polymerization; (**b**) Longitudinal profile of a laser beam and a photopolymerizable region (voxel) formed by two-photon absorption

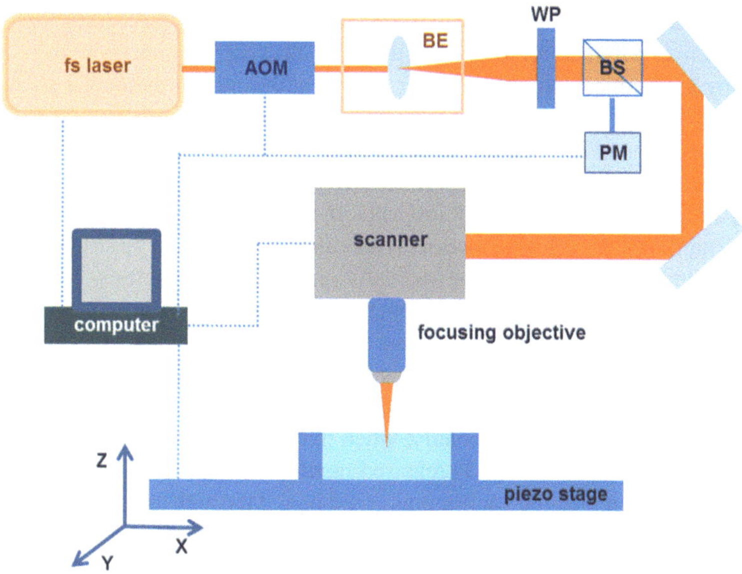

Fig. 3.2 The 2PP setup scheme: *AOM* acousto-optical modulator, *BE* beam expander, *WP* half-wave plate, *BS* beam splitter, *PM* power meter

polarization beam splitter, this mechanism provides a laser power control. Acousto-optical modulator is used as a quick release system, which is needed to turn the laser beam on and off.

A microscope objective focuses laser pulses into a photosensitive material, thereby forming a focal spot, which size depends on the objective type and numerical aperture. The photosensitive material is polymerized in the region of laser focus where the laser intensity is higher than the nonlinear 2PP threshold. Thus, the polymerization region (voxel) can be significantly smaller than the diffraction limit (wavelength). Therefore, structures below 100 nm resolution can be fabricated when using the high magnifying (100×) objective [19].

Precise positioning system and/or a galvanometer scanner are used to move the laser focus within the photosensitive material. The galvanometer scanner allows rapid laser beam movement so that the efficient structuring speed can reach several meters per second. However, the application of galvanometer scanner is limited by the view field of focusing objective. Structures larger than the view field of the objective should be created using a combination of scanner and linear piezoelectric axes [20].

Appropriate positioning system should ensure the accuracy to nm, as well as rapid acceleration and movement at the same time. This dynamic behavior is of great importance for the speed and accuracy of 3D microstructure reproduction. In particular, large (centimeter) range of positioning system movement is important for the quick fabrication of large amount of identical structures for serial studies [21].

During 2PP process, the laser light excites photoinitiator molecules that induce local reactive polymer 3D cross-linking. The polymer photosensitivity is determined by its chemical structure (acrylic groups at the ends of molecular chains) and by the presence of photoinitiators, which induce radicals.

The resolution of 2PP method depends on a number of factors that affect the size of the polymerized voxel: laser focus diameter, wavelength, energy, pulse duration and repetition rate, and pulse peak intensity. In terms of structuring, scanning speed, raster distance, and distance between layers can have an impact on structural resolution. Furthermore, the resolution depends on material properties, photoinitiator type, its concentration, and radical quencher addition [22].

2PP microfabrication technique possesses a unique advantage over other methods. Firstly, the TPA is realized only in a very small 3D volume near the focus (this volume is not greater than $\lambda 3$, where λ—is wavelength). This provides an exceptionally high spatial resolution. Secondly, when materials are exposed to femtosecond laser pulses, the photon energy is released much faster than electrons could pass it to lattice or atom-molecular oscillations could pass it through the photon. Therefore, this excitation is thermally insulating [23]. This feature reduces the probability of autopolymerization process caused by local thermal effects, which provides an ideal optical excitation in the case of photochemical and photophysical reactions.

In addition, many dielectric materials possess the "transparency window" in the red-near-infrared region of spectrum. This "transparency window" does not overlap with any electronic band-band absorption or absorption of atom-molecular vibrations. Typically, the femtosecond laser wavelength falls on the "transparency window" (e.g., 680–1000 nm Ti:sapphire laser). In this regard, the femtosecond laser radiation can penetrate into the interior regions of structures containing material with optical transparency to certain areas and initiate a variety of processes.

3.3 Materials

3.3.1 Photoinitiators

Photoinitiators (Fig. 3.3) are photosensitive compounds, which can absorb one or more photons in a particular energy range that leads to their cleavage and the induction of 3D cross-linking by the formed free radicals, ions, and active fragments. The presence of photoinitiator in photo-curing systems is obligatory for 2PP. These substances should absorb photons at a particular wavelength, which allows the initiation of photoinduced cross-linking.

At the beginning of 2PP method development, most researchers used typical one-photon photoinitiators (e.g., Irgacure 369, Irgacure 184, etc.) because of the absence of commercially available two-photon ones [24, 25] (Table 3.2). These photoinitiators are common in the coating and ink development for metal, wood, and paper production [36–40]. The main mechanisms of their photoinducing activity are monomolecular type I or bimolecular type II reactions [41]. However, these

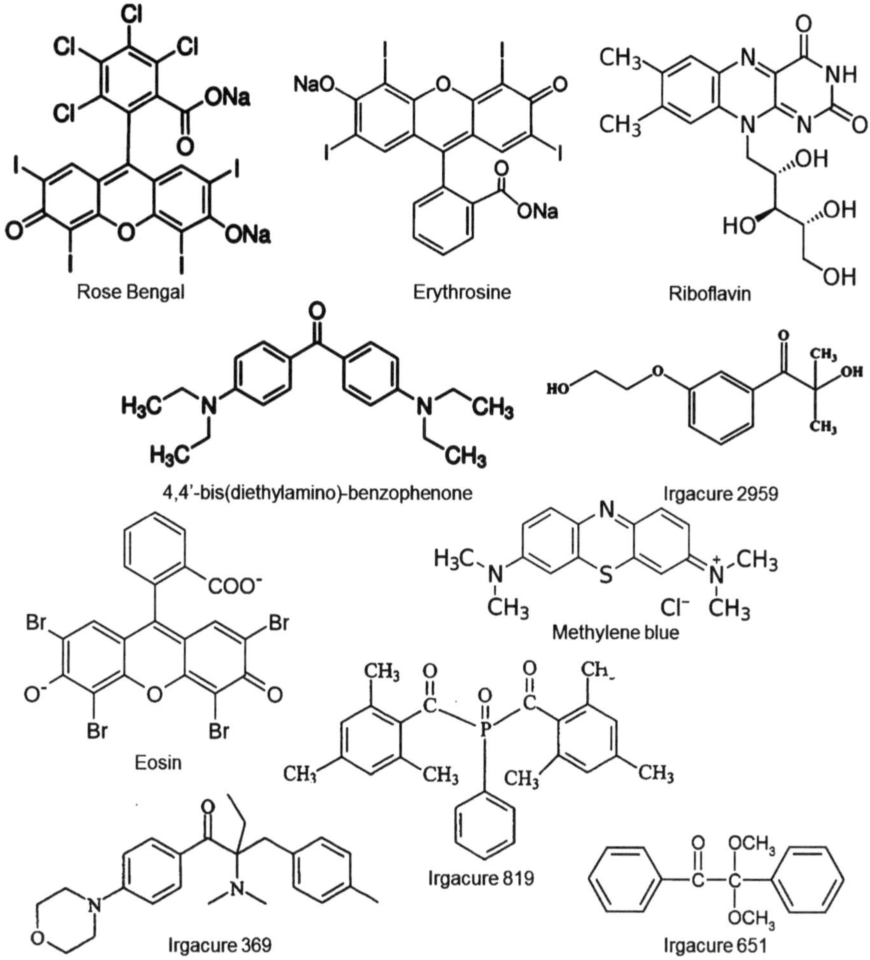

Fig. 3.3 Most common photoinitiators used in 2PP scaffold fabrication

photoinitiators have limited two-photon absorption cross section, which decreases the efficacy of 3D cross-linking initiation.

To date, some researchers have developed two-photon photoinitiators [42]. Nevertheless, most studies include the one-photon photoinitiators like Irgacure 369. In this case, the high radical yield, which is necessary for 3D cross-linking initiation, compensates small two-photon absorption cross section [35, 43].

One of 2PP scaffold applications for TE is based on hydrogels, which require water-soluble (hydrophilic) photoinitiators. Irgacure 2959 is one of the commercially available water-soluble photoinitiators. Due to high biocompatibility, this substance is widely used to fabricate 2PP structures for TE. However, this photoinitiator is not suitable for structuring of polypeptide materials. Since the 2PP

Table 3.2 Photoinitiators which are used in 2PP scaffold fabrication

Photoinitiator	Water solubility	Application in TE	Reference
Irgacure 127	No	Yes	Turunen et al. [26]
Irgacure 369	No	Yes	Nava et al. [27]
Irgacure 651	Yes	No	Jhaveri et al. [28]
Irgacure 819	No	Yes	Heitz et al. [29]
Irgacure 2959	Yes	Yes	Engelhardt et al. [30]
Rose Bengal	Yes	Yes	Engelhardt et al. [30]
Riboflavin, flavin mononucleotide, flavin adenine dinucleotide	Yes	Yes	Kaehr et al. [31], Seidlits et al. [32], Nguyen et al. [33]
4,4'-Bis(diethylamino)-benzophenone	No	Yes	Koroleva et al. [21]
Eosin	Yes	Yes	Huang et al. [34]
Erythrosine	Yes	Yes	Huang et al. [34]
Methylene blue	Yes	Yes	Torgersen et al. [35]

absorption peak of Irgacure 2959 lies at 515–520 nm, irreversible protein denaturation can occur [44].

Hydrophobic photoinitiators can be modified with nonionic surfactants: quaternary ammonium cations, salts of sulfonic and phosphonic acids, and long-chain PEG. Jhaveri et al. [28] showed that Pluronic F-127 increased Irgacure 651 water solubility. However, the main disadvantage of this approach is the need to use high surfactant concentrations to effectively initiate the reaction. High surfactant concentrations increase material cytotoxicity [35].

Up to now, many researchers have applied hydrophilic water-soluble dyes (Rose Bengal, rhodamine, eosin, and methylene blue) and their combinations with amines and other substances (N-methyldiethanolamine, aryl-iodonium, and Omnicat 820). Particularly, some initiating systems for polymerization reactions are based on hydrophilic xanthenes (Rose Bengal, eosin, and erythrosine) and amines as co-initiators. In this system, dye absorbs spontaneously two photons and becomes excited. Intramolecular electron transition with the following proton transition from amine to the excited dye forms highly reactive radicals, which initiate 3D cross-linking. Dyes with co-initiators were used for 2PP structuring of acrylamide [45], pentaerythritol triacrylate [46], and methylated cyclodextrin [47].

3.3.2 Photopolymers

Reactive photo-curing systems (photopolymers) (Table 3.3, Fig. 3.4), which were first used to fabricate porous 2PP scaffolds for TE, were commercially available photoresists (Ormocer®, OrmoComp®, Accura® SI 10, etc.) [26, 48, 49, 70–72]. Successful results of their application opened up new prospects for 2PP in TE.

Table 3.3 Photosensitive materials used in 2PP scaffold fabrication

Photopolymer	Photoinitiator	BD	BC	G	Reference
Ormosil (SZ2080) Organically modified silica	Irgacure 369	No	Yes	No	Nava et al. [27]
Ormosil (SZ2080) Organically modified silica	2-benzyl-2-dimethylamino-1-(4-morpholinophenyl)-butanone-1	No	Yes	No	Malinauskas et al. [48], Danilevičius et al. [49]
Ormocer Organically modified ceramics	Irgacure 369	No	Yes	No	Doraiswamy et al. [50], Doraiswamy et al. [51]
OrmoComp Hybrid polymer	Irgacure 127	No	Yes	No	Turunen et al. [26]
Zr-Si-organic polymer ceramics	4,4′-bis(diethylamino)-benzophenone	No	Yes	No	Terzaki et al. [52], Skoog et al. [53], Koroleva et al. [21], Timashev et al. [54], Vedunova et al. [55]
Methacrylated urethane	Irgacure 369	No	Yes	No	Weiss et al. [56]
Polyethylene glycol diacrylate (258)	4,4′-bis(diethylamino)-benzophenone	No	Yes	Yes	Malinauskas et al. [48], Danilevičius et al. [49]
Polyethylene glycol diacrylate (700)	Irgacure 369	No	Yes	Yes	Weiss et al. [56]
Polyethylene glycol diacrylate (742)	Irgacure 369	No	Yes	Yes	Ovsianikov et al. [57]
Polyethylene glycol diacrylate (742)	Riboflavin, triethanolamine	No	Yes	Yes	Nguyen et al. [33]
Methacrylated branched poly(D,L)-lactide	4,4′-bis(diethylamino)-benzophenone	Yes	Yes	No	Melissinaki et al. [58], Gill et al. [59], Timashev et al. [7]
Methacrylated branched polylactide	Irgacure 369	Yes	Yes	No	Koroleva et al. [16, 17]
Poly(ε-caprolactone-co-trimethylenecarbonate)-b-poly(ethylene glycol)-b-poly(ε-caprolactoneco-trimethylenecarbonate)	4,4′-bis(diethylamino)-benzophenone	Yes	Yes	No	Claeyssens et al. [60]
Pentaerythritol triacrylate, bisphenol A glycidyl methacrylate (80:20)	Irgacure 819	Yes	Yes	No	Heitz et al. [29]
Methacrylated hyaluronic acid	Irgacure 2959	Yes	Yes	Yes	Kufelt et al. [61]
6 wt% allyl-chitosan, 2 wt% methacrylated hyaluronic acid	Irgacure 2959	Yes	Yes	Yes	Timashev et al. [54]

(continued)

Table 3.3 (continued)

Photopolymer	Photoinitiator	BD	BC	G	Reference
Gelatin methacrylamide	Irgacure 2959	Yes	Yes	Yes	Engelhardt et al. [30]
Fibronectin	Rose Bengal	Yes	Yes	Yes	Engelhardt et al. [30]
Bovine serum albumin	Rose Bengal	Yes	Yes	Yes	Pitts et al. ([62], [63])
Bovine serum albumin	Methylene blue	Yes	Yes	Yes	Allen et al. [64], Harper et al. [65]
Bovine serum albumin	Flavin adenine dinucleotide	Yes	Yes	Yes	Kaehr et al. [31], Seidlits et al. [32]
Lyophilized bovine serum albumin	Eosin	Yes	Yes	Yes	Spivey et al. [66]
Lyophilized bovine serum albumin	Methylene blue	Yes	Yes	Yes	Spivey et al. [66]
Collagen	Riboflavin	Yes	Yes	Yes	Kuetemeyer et al. [67]
Collagen	Rose Bengal	Yes	Yes	Yes	Pitts et al. [63], Basu et al. [68]
Fibrinogen	Rose Bengal	Yes	Yes	Yes	Pitts et al. [62], Cunningham et al. [69]

Notes. *BD* biodegradation, *BC* biocompatibility, *G* gelation.

For instance, Marino et al. [71] and Turunen et al. [26] showed that OrmoComp® and SU8 structures supported cell proliferation.

The main idea of scaffold application in TE is temporary cell support. However, materials, like Ormocer®, OrmoComp®, Accura®SI10, and Ormosil (Fig. 3.5), are biostable, and their structure remains the same for a long time. Therefore, most recent studies have focused on biodegradable 2PP materials. The work by Claeyssens et al. [60] is one of the first, and it discusses the synthesis of polycaprolactone derivative with methacrylate groups. The study of its properties showed that this material did not possess cytotoxicity and degraded during tissue formation [60]. Then, other scientists have developed new biodegradable photopolymers based on polylactide, oligolactone, and urethane [7, 56, 58] (Fig. 3.5).

Chemical modification of natural polymers (gelatin, collagen, and hyaluronic acid) is one of the ways to develop biocompatible and biodegradable 2PP photopolymers. These substances in their native state cannot cross-link under laser exposure, even applying their mixture with a high photoinitiator concentration [44]. Therefore, natural proteins and polysaccharides should be modified, for example, with methacrylamide to develop a reactive system. To date, several successful studies have demonstrated this strategy to create biomimetic 2PP photopolymers and fabricate scaffolds from them. One of these materials is biodegradable methacrylamide-modified gelatin. Gelatin is produced by hydrolysis of collagen, which is one of the main ECM components. The gelatin basic structure remains stable under its modification with methacrylamide. Ovsianikov et al. [44] fabricated

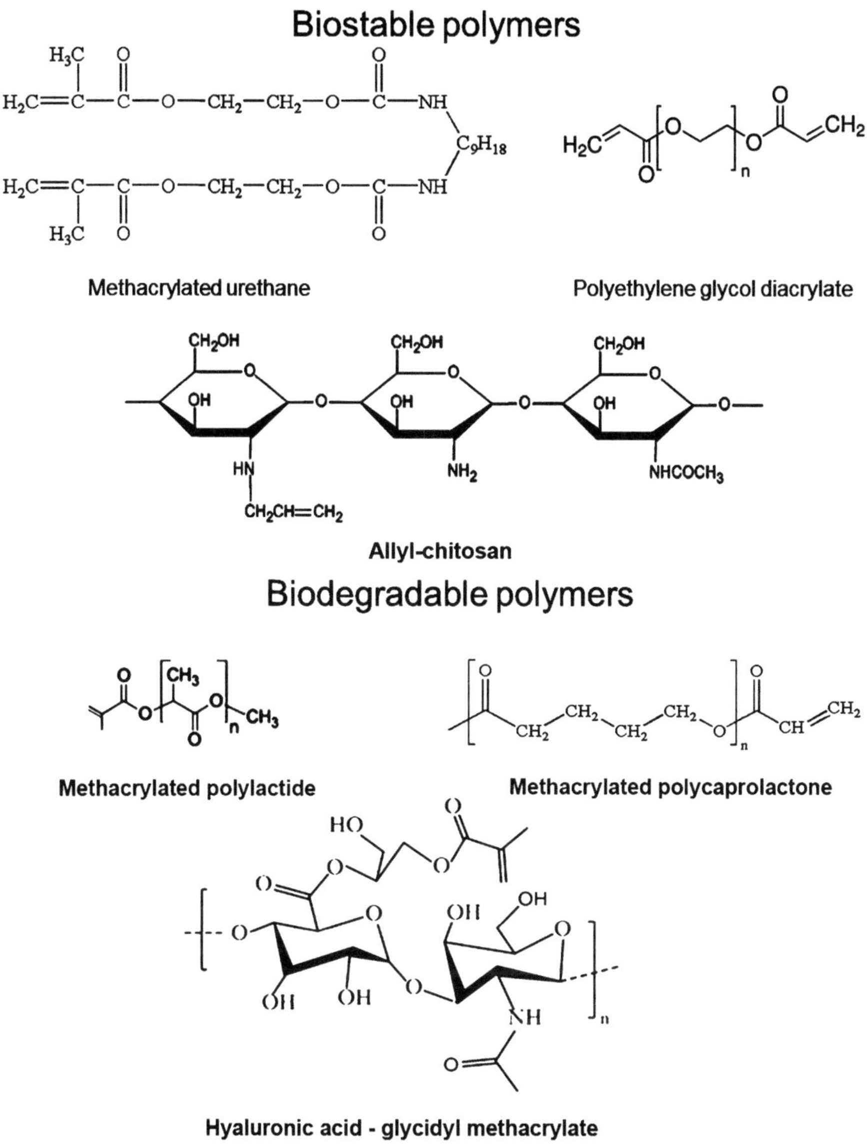

Fig. 3.4 Most common photopolymers used in 2PP scaffold fabrication

biocompatible structures by photocrosslinking of modified gelatin using Irgacure 2959 and under laser exposure at a wavelength 515 nm. These structures promoted stem cell adhesion, proliferation, and differentiation. Kufelt et al. [61] showed that gels from glycidyl methacrylate modified hyaluronic acid possessed no cytotoxicity and no changes in osteoblast and fibroblast metabolic activity. This material was additionally functionalized with epidermal growth factor, which increased significantly the fibroblast proliferation level.

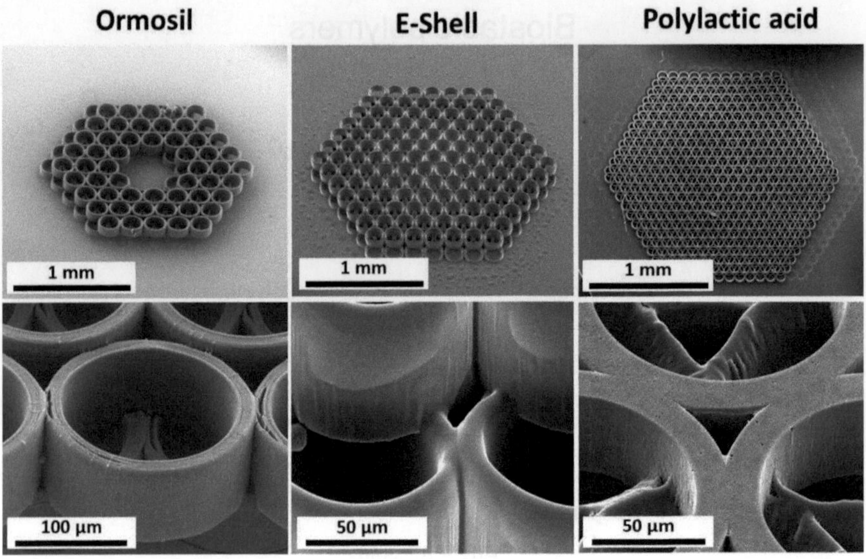

Fig. 3.5 2PP scaffolds from biostable (Ormosil and E-shell) and biodegradable (polylactic acid) materials (authors' data)

Timashev et al. [54] developed gel from allyl chitosan and methacrylated hyaluronic acid. 2PP structures from this material had good biocompatibility. Hippocampal cells, which were cultured on them, had no changes in their metabolic status and formed intercellular contacts.

One of the critical points in 2PP application in TE is material biocompatibility. Polymers and photoinitiators, which were not depleted during polymerization, possess the highest toxicity. Ovsianikov et al. [73] showed that the biggest impact to total material cytotoxicity belongs to photoinitiators. Moreover, various substances had different toxic concentrations. Nevertheless, there is a tendency that higher photoinitiator concentrations lead to higher toxicity levels.

3.4 Applications in Tissue Engineering

3.4.1 Bone Tissue

The bone formation process is complex and still not clearly understood. Due to the bone complexity, one should remember that bone TE relies on three basic factors: ECM/scaffold, cells within matrix/scaffold, and cell environment [74, 75].

To date, bone ECM has been viewed as an inert environment that only provides mechanical stromal support [1]. In fact, ECM is an active and dynamic biochemical system that regulates bone regeneration. It deposits osteogenic and vasculogenic growth factors and provides necessary environment for cell migration, adhesion, proliferation, and differentiation [76, 77].

In bone TE, changes in substrate hardness, nanotopography, microgeometry, and external forces simulate tensile force between cytoskeleton and ECM. For instance, a relatively solid matrix (Young's modulus 25–40 kPa) can induce the osteogenic differentiation of mesenchymal stem cells [78, 79]. Moreover, a rough surface has stronger effects on the osteogenic mesenchymal stem cell differentiation than a smooth surface. This may occur because of stimulation of cell straightening and stretching [80–82].

2PP provides new possibilities in bone TE (Table 3.4). Complex scaffolds from different materials with the required properties can be fabricated using this method (Fig. 3.6). 2PP scaffolds enable the imitation of bone structure and its mechanical properties, the induction of the osteogenic cell differentiation, the deposition and release of biologically active substances and drugs, and the stimulation of bone replacement.

Nowadays, most scientists use biostable and biodegradable photopolymer types in bone TE. For example, the first type includes Zr-Si-based organic polymer ceramics and OrmoComp®, and the second type is presented by methacrylated polylactide (Fig. 3.7) [7, 21, 54, 88]. The common photoinitiator is 4,4′-bis(dimethylamino) benzophenone at a concentration of 1–2%.

Marino et al. [71] received good results with OrmoComp® trabecula-like scaffolds and named them "Osteoprint." They showed that these scaffolds, which were seeded with SaOS-2 human osteosarcoma cells, influenced cell behavior and stimulated their osteogenic differentiation. Marino et al. revealed the upregulation of genes involved SaOS-2 cell development and the increase in hydroxyapatite synthesis and accumulation. Furthermore, in vitro experiments showed that piezoelectric barium titanate nanoparticles, added to OrmoComp®, stimulated the SaOS-2 cell differentiation process [91].

According to Nava et al. [27], human mesenchymal stromal cells, which were cultured on Ormosil 2PP scaffolds, retained their clonogenic potential better than cells on glass. In addition, expression levels of RUNX2 gene and bone sialoprotein were significantly higher in the first case than in the second.

Heitz et al. [29] showed that progenitor cells on 2PP scaffolds with quadratic pores (35 μm) from two acrylate monomers had high level of calcium-binding protein (e.g., collagen type I) synthesis and an increased activation of actin cytoskeleton.

Koroleva et al. [21] revealed that Zr-Si organic-inorganic 2PP scaffolds promoted the spontaneous differentiation of mesenchymal stem cells toward osteogenic lineage without external stimulation. At day 28, cells produced calcium phosphate spherical granules (5–30 μm) both in pores and on a scaffold surface.

However, one of the most promising scaffold materials is polylactides, which are biodegradable and biocompatible. They allow the modeling of any surface nanotopology. Timashev et al. [7] showed that polylactide 2PP scaffolds promoted the in vitro osteogenic differentiation of adipose-derived mesenchymal stem cells. Moreover, the mineralized tissue formation increased significantly after transplantation of these scaffolds, with and without cells, into the cranial bone defect. Bone tissue mineralization, i.e., inorganic substance (particularly, calcium phosphate)

Table 3.4 2PP scaffolds applications in TE

Photopolymer	Photoinitiator	Cell type	Reference
Bone tissue			
Methacrylated branched poly(D,L)lactide	1% 4,4′-bis(dimethylamino) benzophenone	Human adipose-derived stem cells in osteogenic medium (hASCs)	Timashev et al. [7], Shashkova et al. [83]
Zr-Si-based organic polymer ceramics	1% 4,4′-bis(dimethylamino) benzophenone	Human bone marrow-derived stem cells (hBMSCs) and hASCs in osteogenic medium	Skoog et al. [53], Koroleva et al. [21]
Polypyrrole, polyurethane (1:1)	-	MG-63 human osteosarcoma cells	Mihailescu et al. [84]
Zr-Si-based organic polymer ceramic	1% 4,4′-bis(dimethylamino) benzophenone	MC3T3-E1 mouse osteoblastic cells	Terzaki et al. [52]
OrmoComp Hybrid polymer	–	SaOS-2 human osteosarcoma cells	Marino et al. [71]
Ormosil (SZ2080) Organically modified silica	1% Irgacure 369	hBMSCs	Nava et al. [27]
Pentaerythritol triacrylate (PETA)/ bisphenol A-glycidyl methacrylate (BisGMA) (80:20)	2% Irgacure 819	Primary fibroblast cells from human hair roots	Heitz et al. [29]
Zr-Si-based organic polymer ceramic with physically adsorbed and covalently bonded recombinant human bone morphogenetic protein-2 (rhBMP-2)	1% 4,4′-bis(dimethylamino) benzophenone	Mesenchymal stem cells (MSCs)	Chatzinikolaidou et al. [85]
Neural tissue			
Methacrylated branched polylactide	2% Irgacure 369	Primary Schwann cells from adult male Wistar rats	Koroleva et al. [16], [17], [86]
Methacrylated branched polylactide	2% 4,4′-bis(dimethylamino) benzophenone	PC12 and NG108–15 neuronal cells	Melissinaki et al. [58]
Methacrylated branched polylactide	2% 4,4′-bis(dimethylamino) benzophenone	NG108–15 neuronal cells, human neural progenitor cells	Gill et al. [59]

(continued)

Table 3.4 (continued)

Photopolymer	Photoinitiator	Cell type	Reference
Ormocer Organically modified ceramics	Irgacure 369	B35 neuroblast cells	Doraiswamy et al. [50]
OrmoComp Hybrid polymer	–	PC12 neuronal cells, SH-SY5Y human neuroblastoma cells	Marino et al. [87]
OrmoComp Hybrid polymer	2% Irgacure 127	Neuronal cells derived from Regea 08/023 and Regea 06/040 human embryonic cells	Turunen et al. [26]
Zr-Si-based organic polymer ceramics	1% 4,4′-bis(dimethylamino) benzophenone	Hippocampal cells from C57BL/6 J mice embryos	Timashev et al. [88], Vedunova et al. [55]
6 wt.% allyl-chitosan, 2 wt.% methacrylated hyaluronic acid	0.5% Irgacure 2959	Hippocampal cells from C57BL/6 J mice embryos	Timashev et al. [54]
Liver			
Accura SI10	–	Primary hepatocytes from adult male Wistar rats	Hsieh et al. [70]
Cartilaginous tissue			
Gelatin methacrylamide	0.7% Irgacure 2959	Primary chondrocytes from porcine knee	Engelhardt et al. [30]
Fibronectin	4 mM rose Bengal	Primary chondrocytes from porcine knee	Engelhardt et al. [30]
Bovine serum albumin	3 mM rose Bengal	Primary chondrocytes from porcine knee	Engelhardt et al. [30]
Protein-polymer hybrid systems	1% Irgacure 184 or 0.5% EPI	Primary chondrocytes from porcine knee	Engelhardt et al. [30]

(continued)

Table 3.4 (continued)

Photopolymer	Photoinitiator	Cell type	Reference
Methacrylated urethane, polyethylene glycol diacrylate (PEG-DA-700), methacrylated oligolactones (OLMA1, OLMA2)	Irgacure 369	Primary chondrocytes from bovine knee	Weiss et al. [56]
Ormocer	–	Primary bovine chondrocytes	Weiss et al. [72]
Muscular tissue			
AKRE37 Acrylate-based polymer	1–2% thioxanthene-9-one, 4,4′-bis(dimethylamino) benzophenone	Myogenic stem cells from adult rabbit	Malinauskas et al. [48], Danilevičius et al. [49]
Ormosil Organically modified silica	1–2% 2-benzyl-2-dimethylamino-1-(4-morpholinophenyl)-butanone-1	Myogenic stem cells from adult rabbit	Malinauskas et al. [48], Danilevičius et al. [49]
Polyethylene glycol diacrylate (PEG-DA-258)	1–2% 4,4′-bis(dimethylamino) benzophenone	Myogenic stem cells from adult rabbit	Malinauskas et al. [48], Danilevičius et al. [49]
Polyethylene glycol diacrylate (PEG-DA-742)	2% Irgacure 369	Ovine smooth muscle cells	Ovsianikov et al. [57]
Connective tissue			
Poly(ε-caprolactone-co-trimethylenecarbonate)-b-poly(ethylene glycol)-b-poly(ε-caprolactone-co-trimethylenecarbonate)	2% 4,4′-bis(dimethylamino) benzophenone	3T3 NIH fibroblasts	Claeyssens et al. [60]
Ormocer, OrmoDev Organically modified ceramics	–	3T3 NIH fibroblasts	Hidai et al. [89], Jeon et al. [90]
Methacrylated urethane, polyethylene glycol diacrylate (PEG-DA-700), methacrylated oligolactones (OLMA1, OLMA2)	Irgacure 369	3T3 NIH fibroblasts	Weiss et al. [56]
Epithelial tissue			
Ormocer Organically modified ceramics	Irgacure 369	HT1080 epithelial-like cells	Doraiswamy et al. [50]
Ormocer, OrmoDev Organically modified ceramics	–	MDCK epithelial cells	Hidai et al. [89]
Ormocer Organically modified ceramics	1.8% Irgacure 369	Neonatal human epidermal keratinocytes	Doraiswamy et al. [51]

(continued)

Table 3.4 (continued)

Photopolymer	Photoinitiator	Cell type	Reference
Endothelial tissue			
E-Shell 300 (polyacrylics), fibrin	–	HPMEC ST1.6R human pulmonary endothelial cells	Koroleva et al. [16], [17]
Polyethylene glycol diacrylate (PEG-DA-742)	2% Irgacure 369	Ovine primary endothelial cells	Ovsianikov et al. [57]

Fig. 3.6 Key elements in scaffold design for bone TE (by [75])

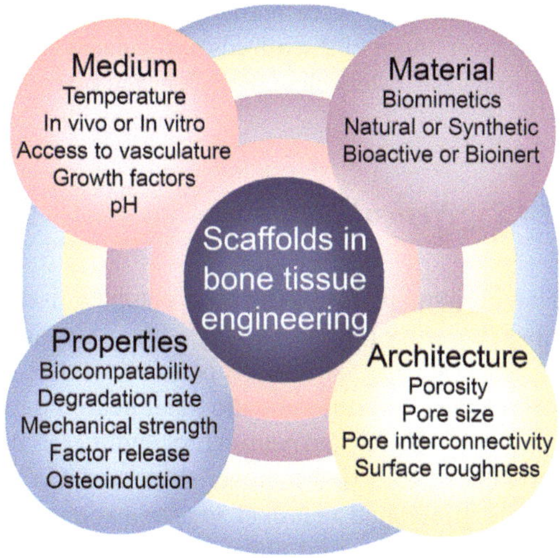

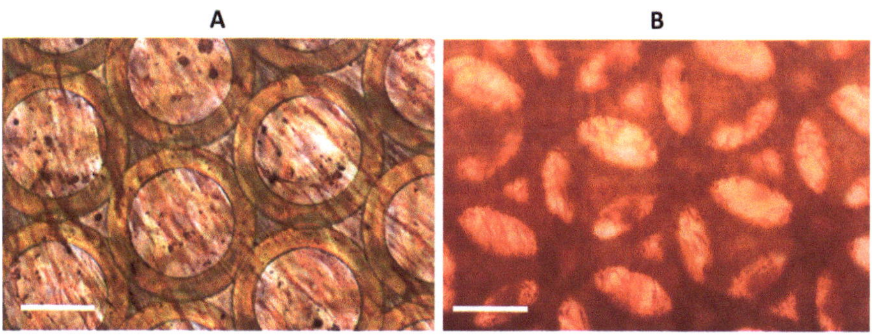

Fig. 3.7 Bone mineralization of hASCs grown on scaffolds from Ormosil (**a**) and polylactide (**b**) in osteogenic medium (Alizarin red S staining; scale bar, 150 μm) (authors' data)

deposition, is the most important criterion of bone formation in a defected place. Cowan et al. [92] showed that on scaffolds the mineralization level of defected places depended strongly on the presence of inoculated cells. By the 10th week, cell-seeded scaffolds demonstrated significantly higher mineralization percentage (92.3%) than scaffolds without cells (65.8%). However, Timashev et al. [7] revealed that polylactide 2PP scaffolds, with and without seeded cells, had good mineralization level by the 10th week. The received results were higher than in other samples. Thus, these polylactide 2PP scaffolds could induce bone regeneration even without cells.

Chatzinikolaidou et al. [85] achieved interesting results. Via covalent binding and physical adsorption, they immobilized recombinant human bone morphogenetic protein 2 (rhBMP-2) onto Zr-Si-based ceramic 2PP scaffolds. Chatzinikolaidou et al. reached high cell biological activity profiles and a sustained and stable rhBMP-2 release. The collagen and calcium levels in cells, cultured on the functionalized scaffolds, were higher than in those cultured in the osteogenic medium. Moreover, these 2PP scaffolds upregulated significantly gene expression of alkaline phosphatase, osteocalcin, RUNX2, and osteonectin.

3.4.2 Neural Tissue

Neural tissue is one of the most complex tissues, and its morphological and functional recovery is an increasingly important area in regenerative medicine.

Substrate mechanical properties together with biochemical factors influence neural stem cell differentiation and neuron development. For instance, mesenchymal stem cells formed neuron progenitor cells on soft substrates (0.1–1 kPa Young's modulus) [93–95]. Moreover, Yim and Leong [96] showed that nano-gratings highly stimulated neuronal marker expression in mesenchymal stem cells compared to retinoic acid alone in unpatterned controls. Scaffolds with filamentous structure promoted differentiation of neural stem cells into motor neurons and axon branching [97].

2PP allows the fabrication of substrates with the required mechanical properties for cell cultivation. In neural TE, both biostable and biodegradable photopolymers are also used (Table 3.4).

One of the most successful studies with biostable polymers in neural TE is the work by Timashev et al. [88], who studied scaffolds from Zr-Si-organic polymer ceramics using murine embryonic neural cells. Dissociated hippocampal cells adhered to a structure surface, formed axons, elongated, and differentiated into adult neurons.

Moreover, Marino et al. [87] revealed that a submicrometric ridge from OrmoComp® promoted the differentiation of rat PC12 neuron-like cells and human SH-SY5Y neural cells. The ridge which consisted of parallel bands with a 2.5 μm space possessed higher positive effects on cells than flat and other sample substrates. Turunen et al. [26] designed special neurocages from OrmoComp® to confine neurons. Although their hypothesis was not proved, cells tended to move toward these structures. In the study by Doraiswamy et al. [50], B35 neuroblast-like cells, which were cultured on 2PP structures from Ormocer®, had good viability and proliferation rates.

In neural TE, most researchers use chemically modified biodegradable polylactides to fabricate scaffolds via 2PP. For example, Koroleva et al. [16, 17] showed that polylactide scaffolds possessed high biocompatibility. SH-SY5Y human neuronal cells had low DNA damage, and primary rat Schwann cells adhered well to their surface, had bipolar and tripolar morphology, and formed intercellular contacts. Furthermore, Melissinaki et al. [58] revealed that the polylactide scaffolds had no significant decrease in the cell viability and proliferation ability of PC12 neuron-like cells and NG108–15 neuroblastoma cells. Gill et al. [59] developed a polylactide microdevice using 2PP to study neuronal stem and neuronal cells in 3D. They showed that these polylactide substrates caused the increase in axon length and cell proliferation in NG108–15 neuroblastoma cell and human neuronal progenitor cell cultures.

Biodegradable hydrogels are promising in 2PP scaffold fabrication. Timashev et al. [54] structured hydrogel from chitosan derivatives and methacrylated hyaluronic acid via 2PP and revealed its compatibility with primary hippocampal cells. They formed a morphologically normal neural network, and their metabolic status was similar to that in the control culture.

3.4.3 Other Tissues

2PP scaffolds are also used in the engineering of cartilaginous, connective, and muscular tissues (Table 3.4). However, most researchers have studied material biocompatibility and cytotoxicity using different cell types, optimized technical approaches, and tried to solve fundamental problems [50, 51, 56, 57, 60, 89, 90]. For instance, Doraiswamy et al. [50, 51] assessed the influence of transdermal drug delivery systems on epithelial cells. Very few works discuss the possibility of 2PP scaffold applications in the engineering of these tissues.

One of these works is the study by Engelhardt et al. [30], who investigated 2PP hybrid polymer-protein lines seeded with primary chondrocytes. Cells cultured on these structures were flat, spindle-shaped, and parallel to the fabricated lines. Mačiulaitis et al. [98] carried out successful in vitro and in vivo experiments with 2PP scaffolds from hybrid organic-inorganic sol–gel photopolymer Ormosil (SZ2080). In the first case, they seeded 3D structures with allogenic rabbit chondrocytes; in the second case, they implanted them into rabbits and examined ex vivo. The biocompatibility of Ormosil scaffolds was high and similar to the control materials (collagen membranes).

To date, only Hsieh et al. [70] have used hepatocytes as a test culture. They fabricated microporous cubic scaffolds from commercial photopolymer Accura SI10 and seeded them with primary hepatocytes derived from male Wistar rats. The colorimetric assays of albumin synthesis and ammonium chloride conversion showed that cells cultured on 3D structures had high functional activity.

Malinauskas et al. [48] and Danilevičius et al. [49] revealed the 2PP scaffold myogenic potential. Biostable scaffolds from AKRE37, Ormosil (SZ2080), Ormocer, and polyethylene glycol diacrylate (PEG-DA-258) were biocompatible and induced myogenic stem cell differentiation.

3.4.4 In Vivo Visualization

In vivo scaffold degradation is one of the main points in TE. Although biostable materials, like OrmoComp®, Ormosil, and OrmoDev [21, 26, 27, 50, 52, 53, 71, 87], are common, there is an increasing tendency to develop and apply biodegradable polymers. Therefore, the estimation of in vivo scaffold degradation rates is crucial for TE, especially, during scaffold implantation and preclinical and clinical trials.

There are many studies on the in vitro estimation of scaffold degradation kinetics [99–101]. However, in vivo assessment of this parameter remains a problem. Most common methods are based on weight changes and histological analysis [102, 103]. A new in vivo noninvasive visualization method is needed to reduce the number of surgical procedures. There are several approaches to solving this problem: biopolymer cross-linking with fluorescent dyes or nanoparticles, fluorophore encapsulation into material, and development of autofluorescent polymers [4, 5, 104–109].

For instance, Zhang et al. [109] used gels from hyaluronic acid linked to fluorescent dye IRDye® 800CW. They introduced them into bone marrow defects in C57Bl/6 J mice. Kim et al. [107] also presented a noninvasive fluorescence method for scaffold imaging. They used powder from porcine cartilage-derived ECM labeled with Cy3 fluorescent dye. In real time, they analyzed the decrease in fluorescence intensity, which depended on the Cy3-labeled scaffold degradation. Kim et al. [105] developed a near-infrared fluorescent scaffold from small intestinal porcine submucosa conjugating with ZW800–1.

However, autofluorescent biodegradable materials are the most attractive [106, 110]. This approach requires no use of additional fluorescent agents, which cannot be homogeneously distributed within scaffolds. This leads to incorrect degradation kinetics values. Moreover, fluorescent agents are toxic and increase the burden on normally functioning organs and tissues (liver, kidney, spleen, etc.).

2PP enables the fabrication of autofluorescent biodegradable scaffolds, which can be observed in real-time via in vivo noninvasive methods (Fig. 3.8). Timashev et al. [7]) used successfully this approach in their study. Polylactide 2PP scaffolds possessed autofluorescence and were implanted after cell seeding into C57/B16 mice. They assessed the in vivo degradation rates due to changes in fluorescence intensity in real time.

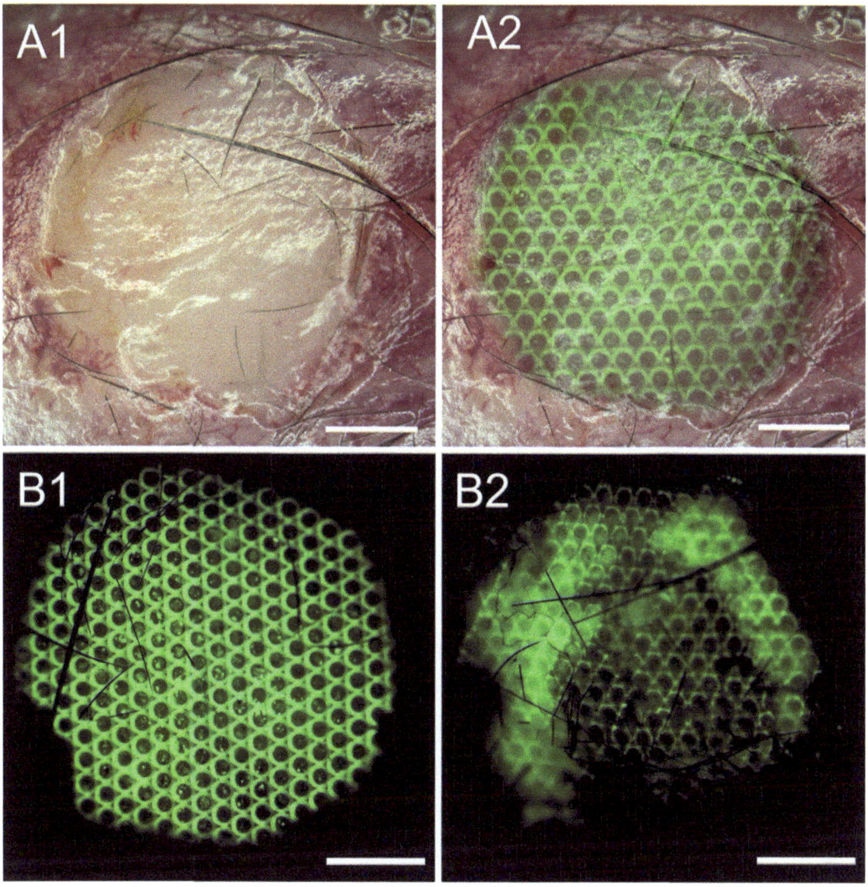

Fig. 3.8 (**a**) 2PP scaffold after one day of its implantation into a mouse skull (A1) and its in vivo visualization using autofluorescence (A2) (scale bar, 1 mm); (**b**) In vivo degradation of a cell seeded autofluorescent polylactide scaffold after implantation (B1) and 5 weeks later (B2) (scale bar, 1 mm) (authors' data)

3.5 Conclusion

The rapid development of 2PP in TE is mostly connected to the possibility of fabricating scaffolds, which are from different materials, have complex architecture, mimic tissue structure and mechanical properties, and induce cell differentiation (Fig. 3.9).

One of the main trends in this development is the use of chemically modified natural polymers with high biocompatibility and tunable biodegradation. To date, new methods of synthesis and analysis [7, 30, 54, 61] have made possible to modify gelatin, hyaluronic acid, chitosan, and others with acrylate and other groups, which are highly reactive and crucial for 2PP process. These modified substances permit

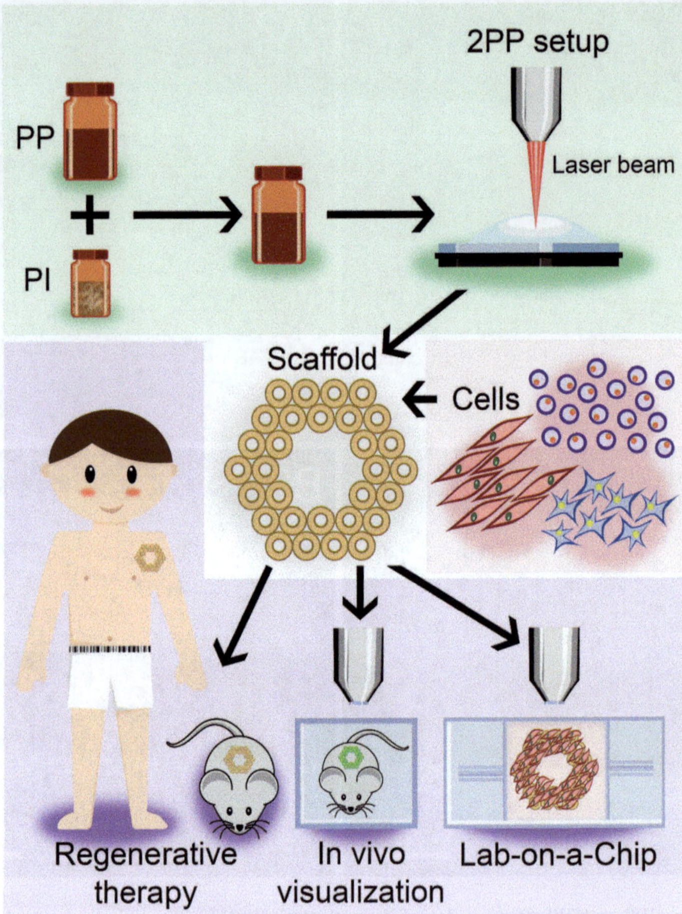

Fig. 3.9 Fabrication of 2PP scaffolds and their possible applications: *PP* photosensitive polymer, *PI* photoinitiator

the 2PP fabrication of scaffolds, which have unique properties which are essential for TE and uncommon for scaffolds from other materials. Therefore, many research groups mainly focus their study on approaches of natural polymer chemical modification.

Moreover, there is a high increase in the development of water-soluble highly efficient photoinitiators, which can initiate 2PP process in photo-curing systems under laser exposure [35, 41]. Further research in this field enable the reduction of photoinitiator concentration in photo-curing systems and the decrease in material cytotoxicity because of this.

One of the new and promising trends in the development of 2PP in TE is the functionalization of 2PP scaffolds using, for instance, nanoparticles or proteins. However, there are only few works [85, 91]. Nevertheless, their number will grow because the 2PP structure functionalization enables the deposit and control of the release of biologically active compounds and drugs. This leads to the increase in the defect tissue and organ replacement during regenerative therapy (Fig. 3.9).

The development of lab-on-a-chip systems and the in vivo visualization of implant degradation are interesting aspects of 2PP application in TE (Fig. 3.9). In the first case, 2PP scaffolds allow the creation of diminutive tissue and organ 3D models, which can be used to study drug pharmacodynamics. This approach is especially attractive for new drug development, because experiments can be carried out quickly. In the second case, 2PP scaffolds can be fabricated from autofluorescent biodegradable polymers that allow the observation of real-time in vivo degradation without adding toxic dyes to their content.

Despite all the achievements of 2PP in TE, there are still many problems to introduce it into medical practice. To date, only several research teams have carried out experiments with primary cell cultures [16, 17, 29, 30, 54, 57, 72, 86] and reached preclinical trials [7, 98].

Thus, 2PP in TE is a promising method, which provides more possibilities of developing tissue-engineered constructs for their implantation into tissue and organ defects, lab-on-a-chip systems, and visual estimation of real-time in vivo material degradation. However, these are still the preliminary stages, and more research is needed to translate its use into clinical practice.

This work was supported by the Russian Science Foundation, grant 16-15-00042

References

1. A.C. Allori, A.M. Sailon, S.M. Warren, Biological basis of bone formation, remodeling, and repair-part I: Biochemical signaling molecules. Tissue Eng. Part B Rev. **14**(3), 259–273 (2008)
2. J.P. Santerre, K. Woodhouse, G. Laroched, et al., Understanding the biodegradation of polyurethanes: From classical implants to tissue engineering materials. Biomaterials **26**, 7457–7470 (2005)
3. H. Li, K. Xue, N. Konga, et al., Silicate bioceramics enhanced vascularization and osteogenesis through stimulating interactions between endothelia cells and bone marrow stromal cells. Biomaterials **35**, 3803–3818 (2014)
4. W. Liu, J. Wei, Y. Chen, et al., Electrospinning of poly(L-lactide) nanofibers encapsulated with water-soluble fullerenes for bioimaging application. ACS Appl. Mater. Interfaces **5**(3), 680–685 (2013a)
5. Y. Liu, J. Lim, S.-H. Teoh, Review: Development of clinically relevant scaffolds for vascularised bone tissue engineering. Biotechnol. Adv. **31**(5), 688–705 (2013b)
6. M.M. Nava, M.T. Raimondi, R. Pietrabissa, Controlling self-renewal and differentiation of stem cells via mechanical cues. J Biomed Biotechnol **2012**, 1 (2012). https://doi.org/10.1155/2012/797410
7. P. Timashev, D. Kuznetsova, A. Koroleva, et al., Novel biodegradable star-shaped polylactide scaffolds for bone regeneration fabricated by two-photon polymerization. Nanomedicine **11**, 1041 (2016a). https://doi.org/10.2217/nnm-2015-0022
8. P.J.S. Bártolo, H. Almeida, Rapid prototyping and manufacturing for tissue engineering scaffolds. Int J Comput Appl Technology **36**(1), 1–9 (2009)

9. M.T. Raimondi, S.M. Eaton, M.M. Nava, et al., Two-photon laser polymerization: From fundamentals to biomedical application in tissue engineering and regenerative medicine. J. Appl. Biomater. Biomech. **10**(1), 55–65 (2012)
10. S. Farid, S. Shirazi, S. Gharehkhani, et al., A review on powder-based additive manufacturing for tissue engineering: Selective laser sintering and inkjet 3D printing. Sci. Technol. Adv. Mater. **16**, 033502 (2015)
11. K. Singh, B.R. Singh, H. Singh, Development of rapid tooling using fused deposition modeling: A review. Rapid Prototyping J. **22**(2), 281 (2016). https://doi.org/10.1108/RPJ-04-2014-0048
12. S.V. Murphy, A. Atala, 3D bioprinting of tissues and organs. Nat. Biotechnol. **32**, 773 (2014). https://doi.org/10.1038/nbt.2958
13. Carvalho C (2017) .There's only one 3D–bioplotter. https://envisiontec.com/3d-printers/3d-bioplotter/. Accessed 26 Apr 2017
14. BioBot (2017) What will you build. https://www.biobots.io/biobot-1/. Accessed 26 Apr 2017
15. M. Malinauskas, M. Farsari, A. Piskarskas, et al., Ultrafast laser nanostructuring of photopolymers: A decade of advances. Phys. Rep. **533**(1), 1–31 (2013)
16. A. Koroleva, A.A. Gill, I. Ortega, et al., Two-photon polymerization-generated and micromolding-replicated 3D scaffolds for peripheral neural tissue engineering applications. Biofabrication **4**, 025005 (2012a)
17. A. Koroleva, S. Gittard, S. Schlie, et al., Fabrication of fibrin scaffolds with controlled microscale architecture by a two-photon polymerization–micromolding technique. Biofabrication **4**, 015001 (2012b)
18. J. Serbin, A. Egbert, A. Ostendorf, et al., Femtosecond laser-induced two-photon polymerization of organic-inorganic hybrid materials for applications in photonics. Opt. Lett. **28**(5), 301–303 (2003)
19. M. Emons, K. Obata, T. Binhammer, et al., Two-photon polymerization technique with sub-50 nm resolution by sub-10 fs laser pulses. Opt. Mater. Express **2**(7), 942–947 (2012)
20. A. Ovsianikov, A. Ostendorf, B. Chichkov, Three-dimensional photofabrication with femtosecond lasers for applications in photonics and biomedicine. Appl. Surf. Sci. **253**(15), 6599–6602 (2007)
21. A. Koroleva, A. Deiwick, A. Nguyen, et al., Osteogenic differentiation of human mesenchymal stem cells in 3-D Zr-Si organic- inorganic scaffolds produced by two-photon polymerization technique. PLoS One **10**(2), e0118164 (2015)
22. F. Jipa, M. Zamfirescu, A. Velea, et al., Femtosecond laser lithography in organic and non-organic materials, in *Updates in Advanced Lithography*, ed. by S. Hosaka (Ed), (InTech, 2013). https://doi.org/10.5772/56579
23. P. Saeta, J.K. Wang, Y. Siegal, et al., Ultrafast electronic disordering during femtosecond laser melting of GaAs. Phys. Rev. Lett. **67**, 1023–1026 (1991)
24. C. Heller, N. Pucher, B. Seidl, et al., One- and two-photon activity of cross-conjugated photoinitiators with bathochromic shift. J. Polym. Sci. A Polym. Chem. **45**, 3280–3291 (2007)
25. S. Maruo, O. Nakamura, S. Kawata, Three-dimensional microfabrication with two-photon-absorbed photopolymerization. Opt. Lett. **22**(2), 132–134 (1997)
26. S. Turunen, E. Käpylä, M. Lähteenmäki, et al., Direct laser writing of microstructures for the growth guidance of human pluripotent stem cell derived neuronal cells. Opt. Lasers Eng. **55**, 197–204 (2014)
27. M.M. Nava, N. Di Maggio, T. Zandrini, et al., Synthetic niche substrates engineered via two-photon laser polymerization for the expansion of human mesenchymal stromal cells. J. Tissue Eng. Regen. Med. **11**, 2836 (2016). https://doi.org/10.1002/term.2187
28. S.J. Jhaveri, J.D. McMullen, R. Sijbesma, et al., Direct three-dimensional microfabrication of hydrogels via two-photon lithography in aqueous solution. Chem. Mater. **21**(10), 2003–2006 (2009)
29. J. Heitz, C. Plamadeala, M. Wiesbauer, et al., Bone-forming cells with pronounced spread into the third dimension in polymer scaffolds fabricated by two-photon polymerization. J. Biomed. Mater. Res. A **105**, 891 (2016). https://doi.org/10.1002/jbm.a.35959

30. S. Engelhardt, E. Hoch, K. Borchers, et al., Fabrication of 2D protein microstructures and 3D polymer–protein hybrid microstructures by two-photon polymerization. Biofabrication **3**, 025003 (2011)
31. B. Kaehr, R. Allen, D.J. Javier, et al., Guiding neuronal development with in situ microfabrication. PNAS **101**(46), 16104–16108 (2004)
32. S.K. Seidlits, C.E. Schmidt, J.B. Shear, High-resolution patterning of hydrogels in three dimensions using direct-write photofabrication for cell guidance. Adv. Funct. Mater. **19**, 3543–3551 (2009)
33. A.K. Nguyen, S.D. Gittard, A. Koroleva, et al., Two-photon polymerization of polyethylene glycol diacrylate scaffolds with riboflavin and triethanolamine used as a water-soluble photoinitiator. Regen. Med. **8**(6), 725–738 (2013)
34. S. Huang, A.A. Heikal, W.W. Webb, Two-photon fluorescence spectroscopy and microscopy of NAD(P)H and flavoprotein. Biophys. J. **82**, 2811–2825 (2002)
35. J. Torgersen, X.-H. Qin, Z. Li, et al., Hydrogels for two-photon polymerization: A toolbox for mimicking the extracellular matrix. Adv. Funct. Mater. **23**, 4542 (2013). https://doi.org/10.1002/adfm.201203880
36. A. Cataldi, A.E. Corcione, M. Frigione, et al., Photocurable resin/microcrystalline cellulose composites for wood protection: Physical-mechanical characterization. Prog. Org. Coat. **99**, 230–239 (2016)
37. J.-S. Choi, J. Seo, S.B. Khanc, et al., Effect of acrylic acid on the physical properties of UV-cured poly(urethane acrylate-co-acrylic acid) films for metal coating. Prog. Org. Coat. **71**, 110–116 (2011)
38. P. Kiruthika, R. Subasri, A. Jyothirmayi, et al., Effect of plasma surface treatment on mechanical and corrosion protection properties of UV-curable sol-gel based GPTS-ZrO2 coatings on mild steel. Surf. Coat. Technol. **204**, 270–1276 (2010)
39. S. Papilloud, D. Baudraz, Migration tests for substrates printed with UV ink systems in aqueous simulants. Prog. Org. Coat. **45**, 231–237 (2002)
40. C. Sow, B. Riedl, P. Blanchet, UV-waterborne polyurethane-acrylate nanocomposite coatings containing alumina and silica nanoparticles for wood: Mechanical, optical, and thermal properties assessment. J. Coat. Technol. Res. **8**, 211–221 (2011)
41. A. Ovsianikov, V. Mironov, J. Stampfl, et al., Engineering 3D cell-culture matrices: Multiphoton processing technologies for biological and tissue engineering applications. Expert Rev. Med. Devices **9**(6), 613–633 (2012)
42. F. Hao, Z. Liu, M. Zhang, et al., Four new two-photon polymerization initiators with varying donor and conjugated bridge: Synthesis and two-photon activity. Spectrochim. Acta A Mol. Biomol. Spectrosc. **118**, 538–542 (2014)
43. K.J. Schafer, J.M. Hales, M. Balu, et al., Two-photon absorption cross-sections of common photoinitiators. J. Photochem. Photobiol. **162**, 497–502 (2004)
44. A. Ovsianikov, A. Deiwick, S. Van Vlierberghe, et al., Laser fabrication of 3D gelatin scaffolds for the generation of bioartificial tissues. Materials **4**, 288–299 (2011a)
45. P.J. Campagnola, D.M. Delguidice, G.A. Epling, et al., 3-dimensional submicron polymerization of acrylamide by multiphoton excitation of xanthene dyes. Macromolecules **33**, 1511–1513 (2000)
46. M. Farsari, G. Filippidis, K. Sambani, et al., Two-photon polymerization of an eosin Y-sensitized acrylate composite. J. Photochem. Photobiol. A Chem. **181**, 132–135 (2006)
47. S. Li, L. Li, F. Wu, et al., A water-soluble two-photon photopolymerization initiation system: Methylated--cyclodextrin complex of xanthene dye/aryliodonium salt. J. Photochem. Photobiol. A Chem. **203**, 211–215 (2009)
48. M. Malinauskas, P. Danilevicius, D. Baltriukiene˙, et al., 3D artificial polymeric scaffolds for stem cell growth fabricated by femtosecond laser. Lithuanian J. Phys. **50**(1), 75–82 (2010)
49. P. Danilevičius, A. Žukauskas, G. Bičkauskaitė, et al., Laser-micro/nanofabricated 3D polymers for tissue engineering applications. Lat. J. Phys. Tech. Sci. **2**, 32–43 (2011)

50. A. Doraiswamy, C. Jin, R.J. Narayan, et al., Two photon induced polymerization of organic–inorganic hybrid biomaterials for microstructured medical devices. Acta Biomater. **2**, 267–275 (2006)
51. A. Doraiswamy, A. Ovsianikov, S.D. Gittard, et al., Fabrication of microneedles using two photon polymerization for transdermal delivery of nanomaterials. J. Nanosci. Nanotechnol. **10**, 6305–6312 (2010)
52. K. Terzaki, M. Kissamitaki, A. Skarmoutsou, et al., Pre-osteoblastic cell response on three-dimensional, organic-inorganic hybrid material scaffolds for bone tissue engineering. J. Biomed. Mater. Res. Part A **101A**, 2283 (2013). https://doi.org/10.1002/jbm.a.34516
53. S.A. Skoog, A.K. Nguyen, G. Kumar, et al., Two-photon polymerization of 3-D zirconium oxide hybrid scaffolds for long-term stem cell growth. Biointerphases **9**(2), 029014 (2014)
54. P.S. Timashev, K.N. Bardakova, N.V. Minaeva, et al., Compatibility of cells of the nervous system with structured biodegradable chitosan-based hydrogel matrices. Appl. Biochem. Microbiol. **52**(5), 508–514 (2016b)
55. M.V. Vedunova, P.S. Timashev, T.A. Mishchenko, et al., Formation of neural networks in 3D scaffolds fabricated by means of laser microstereolithography. Bull. Exp. Biol. Med. **161**(4), 616–621 (2016)
56. T. Weiß, R. Schade, T. Laube, et al., Two-photon polymerization of biocompatible photopolymers for microstructured 3D biointerfaces. Adv. Eng. Mater. **13**(9), 264–273 (2011)
57. A. Ovsianikov, M. Gruene, M. Pflaum, et al., Laser printing of cells into 3D scaffolds. Biofabrication **2**, 014104 (2010)
58. V. Melissinaki, A.A. Gill, I. Ortega, et al., Direct laser writing of 3D scaffolds for neural tissue engineering applications. Biofabrication **3**, 045005 (2011)
59. A.A. Gill, Í. Ortega, S. Kelly, et al., Towards the fabrication of artificial 3D microdevices for neural cell networks. Biomed. Microdevices **17**, 27–37 (2015)
60. F. Claeyssens, E.A. Hasan, A. Gaidukeviciute, et al., Three-dimensional biodegradable structures fabricated by two-photon polymerization. Langmuir **25**(5), 3219–3223 (2009)
61. O. Kufelt, A. El-Tamer, C. Sehring, et al., Hyaluronic acid based materials for scaffolding via two-photon polymerization. Biomacromolecules **15**(2), 650–659 (2014)
62. J.D. Pitts, P.J. Campagnola, G.A. Epling, et al., Submicron multiphoton free-form fabrication of proteins and polymers: Studies of reaction efficiencies and applications in sustained release. Macromolecules **33**, 1514–1523 (2000)
63. J.D. Pitts, A.R. Howell, R. Taboada, et al., New photoactivators for multiphoton excited three-dimensional submicron cross-linking of proteins: Bovine serum albumin and type 1 collagen. Photochem. Photobiol. **76**(2), 135–144 (2002)
64. R. Allen, R. Nielson, D.D. Wise, et al., Catalytic three-dimensional protein architectures. Anal. Chem. **77**, 5089–5095 (2005)
65. J.C. Harper, S.M. Brozik, C.J. Brinker, et al., Biocompatible microfabrication of 3D isolation chambers for targeted confinement of individual cells and their progeny. Anal. Chem. **84**(21), 8985–8989 (2012)
66. E.C. Spivey, E.T. Ritschdorff, J.L. Connell, et al., Multiphoton lithography of unconstrained three-dimensional protein microstructures. Adv. Funct. Mater. **23**(3), 333–339 (2012)
67. K. Kuetemeyer, G. Kensah, M. Heidrich, et al., Two-photon induced collagen cross-linking in bioartificial cardiac tissue. Opt. Express **19**(17), 15996–16007 (2011)
68. S. Basu, V. Rodionov, M. Terasaki, et al., Multiphoton-excited microfabrication in live cells via rose Bengal cross-linking of cytoplasmic proteins. Opt. Lett. **30**(2), 159–161 (2005)
69. L.P. Cunningham, M.P. Veilleux, P.J. Campagnola, Freeform multiphoton excited microfabrication for biological applications using a rapid prototyping CAD-based approach. Opt. Express **14**(19), 8613–8621 (2006)
70. T.M. Hsieh, C.W.B. Ng, K. Narayanan, et al., Three-dimensional microstructured tissue scaffolds fabricated by two-photon laser scanning photolithography. Biomaterials **31**, 7648–7652 (2010)

71. A. Marino, C. Filippeschi, G.G. Genchi, et al., The osteoprint: A bioinspired two-photon polymerized 3-D structure for the enhancement of bone-like cell differentiation. Acta Biomater. **10**, 4304–4313 (2014)
72. T. Weiss, G. Hildebrand, R. Schade, et al., Two-photon polymerization for microfabrication of three-dimensional scaffolds for tissue engineering application. Eng. Life Sci. **9**(5), 384–390 (2009)
73. A. Ovsianikov, M. Malinauskas, S. Schlie, et al., Three-dimensional laser micro- and nano-structuring of acrylated poly(ethylene glycol) materials and evaluation of their cytoxicity for tissue engineering applications. Acta Biomater. **7**, 967–974 (2011b)
74. A.R. Amini, C.T. Laurencin, S.P. Nukavarapu, Bone tissue engineering: Recent advances and challenges. Crit. Rev. Biomed. Eng. **40**(5), 363–408 (2012)
75. C. Szpalski, M. Wetterau, J. Barr, et al., Bone tissue engineering: Current strategies and techniques—Part I: Scaffolds. Tissue Eng. Part B **18**(4), 246–256 (2012)
76. K.H. Choi, B.H. Choi, S.R. Park, et al., The chondrogenic differentiation of mesenchymal stem cells on an extracellular matrix scaffold derived from porcine chondrocytes. Biomaterials **31**, 5355–5365 (2010)
77. K. Narayanan, K.J. Leck, S. Gao, et al., Three- dimensional reconstituted extracellular matrix scaffolds for tissue engineering. Biomaterials **30**, 4309–4317 (2009)
78. A.J. Engler, S. Sen, H.L. Sweeney, et al., Matrix elasticity directs stem cell lineage specification. Cell **126**(4), 677–689 (2006)
79. C.B. Khatiwala, P.D. Kim, S.R. Peyton, et al., ECM compliance regulates osteogenesis by influencing MAPK signaling downstream of RhoA and ROCK. J. Bone Miner. Res. **24**(5), 886–898 (2009)
80. E. Engel, E. Martínez, C.A. Mills, et al., Mesenchymal stem cell differentiation on micro-structured poly (methyl methacrylate) substrates. Ann. Anat. **191**(1), 136–144 (2009)
81. E. Martinez, E. Engel, J.A. Planell, et al., Effects of artificial micro- and nano-structured surfaces on cell behavior. Ann. Anat. **191**, 126–135 (2009)
82. S. Martino, F. D'Angelo, I. Armentano, et al., Hydrogenated amorphous carbon nanopatterned film designs drive human bone marrow mesenchymal stem cell cytoskeleton architecture. Tissue Eng. Part A **15**(10), 3139–3149 (2009)
83. V.T. Shashkova, I.A. Matveeva, N.N. Glagolev, et al., Synthesis of polylactide acrylate derivatives for the preparation of 3D structures by photo-curing. Mendeleev Commun. **26**, 418–420 (2016)
84. M. Mihailescu, R.C. Popescu, A. Matei, Investigation of osteoblast cells behavior in polymeric 3D micropatterned scaffolds using digital holographic microscopy. Appl. Opt. **53**(22), 4850–4858 (2014)
85. M. Chatzinikolaidou, C. Pontikoglou, K. Terzaki, et al., Recombinant human bone morphogenetic protein 2 (rhBMP-2) immobilized on laser-fabricated 3D scaffolds enhance osteogenesis. Colloids Surf. B Biointerfaces **149**(1), 233–242 (2017)
86. A.V. Koroleva, D.S. Guseva, N.A. Konovalov, et al., Polylactide-based biodegradable scaffolds fabricated by two-photon polymerization for neurotransplantation. CTM **8**(4), 23–28 (2016)
87. A. Marino, G. Ciofani, C. Filippeschi, et al., Two-photon polymerization of sub-micrometric patterned surfaces: Investigation of cell-substrate interactions and improved differentiation of neuron-like cells. ACS Appl. Mater. Interfaces **5**, 13012–13021 (2013)
88. P.S. Timashev, M.V. Vedunova, D. Guseva, et al., 3D in vitro platform produced by two-photon polymerization for the analysis of neural network formation and function. Biomed. Phys. Eng. Express **2**(3), 1–8 (2016c)
89. H. Hidai, H. Jeon, D.J. Hwang, et al., Self-standing aligned fiber scaffold fabrication by two photon photopolymerization. Biomed. Microdevices **11**, 643–652 (2009)
90. H. Jeon, E. Kim, C.P. Grigoropoulos, Measurement of contractile forces generated by individual fibroblasts on self-standing fiber scaffolds. Biomed. Microdevices **13**, 107–115 (2011)

91. A. Marino, J. Barsotti, G. de Vito, et al., Two-photon lithography of 3D nanocomposite piezoelectric scaffolds for cell stimulation. ACS Appl. Mater. Interfaces **7**, 25574–25579 (2015)
92. C.M. Cowan, Y.Y. Shi, O.O. Aalami, et al., Adipose-derived adult stromal cells heal critical-size mouse calvarial defects. Nat. Biotechnol. **22**(5), 560–567 (2004)
93. A. Banerjee, M. Arha, S. Choudhary, et al., The influence of hydrogel modulus on the proliferation and differentiation of encapsulated neural stem cells. Biomaterials **30**(27), 4695–4699 (2009)
94. N.D. Leipzig, M.S. Shoichet, The effect of substrate stiffness on adult neural stemcell behavior. Biomaterials **30**(36), 6867–6878 (2009)
95. K. Saha, A.J. Keung, E.F. Irwin, Y. Li, et al., Substrate modulus directs neural stem cell behavior. Biophys. J. **95**(9), 4426–4438 (2008)
96. E.K.F. Yim, S.W. Pang, K.W. Leong, Synthetic nanostructures inducing differentiation of human mesenchymal stem cells into neuronal lineage. Exp. Cell Res. **313**(9), 1820–1829 (2007)
97. L. Binan, C. Tendey, G. De Crescenzo, et al., Differentiation of neuronal stem cells into motor neurons using electrospun poly-L-lactic acid/gelatin scaffold. Biomaterials **35**, 664–674 (2014)
98. J. Mačiulaitis, M. Deveikytė, S. Rekštytė, et al., Preclinical study of SZ2080 material 3D microstructured scaffolds for cartilage tissue engineering made by femtosecond direct laser writing lithography. Biofabrication **7**, 015015 (2015)
99. A.R. Costa-Pinto, A.M. Martins, M.J. Castelhano-Carlos, et al., In vitro degradation and in vivo biocompatibility of chitosan–poly (butylene succinate) fiber mesh scaffolds. J. Bioactive Compatible Polym.: Biomed. Appl. **29**(2), 137–151 (2014)
100. C. de Oliveira Renó, N.C. Pereta, C.A. Bertran, et al., Study of in vitro degradation of brushite cements scaffolds. J. Mater. Sci. Mater. Med. **25**, 2297–2303 (2014)
101. N. Zhu, D. Cooper, X.B. Chen, et al., A study on the in vitro degradation of poly(l-lactide)/chitosan microspheres scaffolds. Front. Mater. Sci. **7**, 76–82 (2013)
102. Y. He, Y. Dong, F. Cui, et al., Ectopic osteogenesis and scaffold biodegradation of nano-hydroxyapatite-chitosan in a rat model. PLoS One **10**(8), e0135366 (2015)
103. S.-H. Park, E.S. Gil, H. Shi, et al., Relationships between degradability of silk scaffolds and osteogenesis. Biomaterials **31**(24), 6162–6172 (2010)
104. J. Kim, K.S. Kim, G. Jiang, et al., In vivo real-time bioimaging of hyaluronic acid derivatives using quantum dots. Biopolymers **89**(12), 1144–1153 (2008)
105. S.H. Kim, J.H. Lee, H. Hyun, et al., Near-infrared fluorescence imaging for noninvasive trafficking of scaffold degradation. Sci. Rep. **3**, 1198 (2013). https://doi.org/10.1038/srep01198
106. J. Yang, Y. Zhang, S. Gautam, et al., Development of aliphatic biodegradable photoluminescent polymers. PNAS **106**(25), 10086–10091 (2009)
107. H.J. Kim, S. Lee, H.-W. Yun, et al., In vivo degradation profile of porcine cartilage-derived extracellular matrix powder scaffolds using a non-invasive fluorescence imaging method. J. Biomater. Sci. Polym. Ed. **27**(2), 177–190 (2016)
108. Y. Zhang, J. Yang, Design strategies for fluorescent biodegradable polymeric biomaterials. J. Mater. Chem. B **1**, 132–148 (2013)
109. Y. Zhang, F. Rossi, S. Papa, et al., Non-invasive in vitro and in vivo monitoring of degradation of fluorescently labeled hyaluronan hydrogels for tissue engineering applications. Acta Biomater. **30**, 188–198 (2016)
110. S. Duan, S. Ma, Z. Huang, et al., Visualization of in vivo degradation of aliphatic polyesters by a fluorescent dendritic star macromolecule. Biomed. Mater. **10**(6), 065003 (2015)

Part II
Applications

Chapter 4
The Use of Photo-Activatable Materials for the Study of Cell Biomechanics and Mechanobiology

Michelle E. Pede and James H. Henderson

4.1 Introduction

The goal of this chapter is to highlight photo-activatable materials that have been used to study cell biomechanics and mechanobiology and to discuss current and potential future applications of these materials. For the purposes of this chapter, a photo-activatable material will be defined as a material that changes its bulk shape, surface chemistry, or surface topography when exposed to electromagnetic energy in the ultraviolet (UV; wavelengths of 10–390 nm), visible (390–700 nm), or near-infrared (NIR; 700–2500 nm) spectra. Although photo-activatable materials are increasingly being designed for applications involving biological systems, our focus here will be intentionally restricted to applications of a biomechanical or mechanobiological nature.

Herein we adopt a working definition of cell biomechanics as the study of how physical forces interact with cells and of cell mechanobiology as the study of how cells respond biologically to mechanical forces. Much can and has been learned by studying the biomechanics and mechanobiology of cells using static (unchanging) materials, such as inert polymeric substrates and scaffolds. Yet, in many living tissues, cells reside in dynamic extracellular matrices (ECM) subject to changing mechanical loads and deformations. These changes in mechanical loading occur as a result of external forces, forces generated by growing tissues, or changes in the material properties (e.g., stiffness, elasticity changes) of the ECM over time. Interactions between cells and their dynamic ECM affect cell behaviors such as gene regulation, proliferation, adhesion, protein and ECM component synthesis, morphology (shape) and orientation, migration, apoptosis (cell death), and stem cell differentiation fate [1] (Fig. 4.1). Therefore, changes in any of the physical forces

M. E. Pede · J. H. Henderson (✉)
Syracuse University, Syracuse, NY, USA
e-mail: mpede@syr.edu; jhhender@syr.edu

J. Van Hoorick et al. (eds.), *Polymer and Photonic Materials Towards Biomedical Breakthroughs*, Micro- and Opto-Electronic Materials, Structures, and Systems, https://doi.org/10.1007/978-3-319-75801-5_4

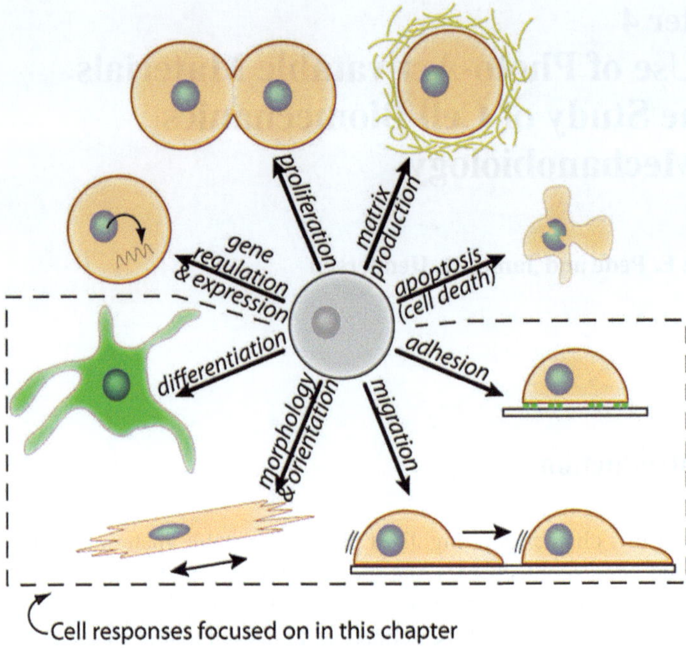

Fig. 4.1 Cell responses to physical forces that are often studied in cell biomechanics and mechanobiology. Note that the responses included in the dashed box are those that will be highlighted in this chapter

exerted on a cell or its ECM may affect these cell behaviors. Additionally, cells can, in turn, remodel their surrounding ECM in response to changes in these physical forces. For example, it is well known that bone becomes stiffer and stronger under increased compressive loads [2]. The dynamic nature of many in vivo mechanobiological environments has motivated the development of dynamic materials for the study of cell biomechanics and mechanobiology. These materials are being used to create dynamic in vitro environments through which researchers are gaining a better understanding of development, disease progression, and wound healing and applying those findings in the development of new and improved clinical treatments.

With this goal of more meaningfully studying and understanding cell biomechanics and mechanobiology, a move has been made from traditional, static techniques to new techniques that better replicate the dynamic environment of natural ECM. Early studies of cell biomechanics and mechanobiology employed mechanical devices that applied loads directly to cells or the substrates to which they were attached. One of the earliest methods developed for measuring the mechanical properties of individual cells came from Mitchison and Swann, who approximated cell stiffness by applying nanonewton-scale forces to the surface of sea urchin eggs and measuring the resulting deformations [3]. This technique, known as micropipette aspiration, and others such as atomic force microscopy (AFM) and optical trapping, have been used to measure a number of cell mechanical properties including individual cell stiffness and attachment strength [4–9]. Other devices have been developed to investigate the collective response of cultured cells and tissues to mechanical stimulation. For example, compression chambers have demonstrated that chondrocytes increase proteoglycan synthesis in response to cyclic, high-frequency loading [10], and flow chambers have shown that endothelial cells reorient their cytoskeleton to align in the flow direction [11]. All of these methods for studying cell biomechanics and mechanobiology, however, require cumbersome devices and/or direct contact with cells or substrates.

In recent years, there has been a move to develop "smart" biomaterials that shift the mechanical functionality of mechanobiology platforms away from external devices and into the material itself. These materials possess properties that can be actively changed upon exposure to an external stimulus such as temperature [12], pH [13, 14], electric [15] or magnetic [16] fields, hydration [17], or, of particular interest in this chapter, light [18–21]. Light-responsive materials are an emerging portion of this field and are increasingly being applied to biomechanics and mechanobiology. One main advantage of utilizing photo-irradiation as the external stimulus is that it allows for remote activation that can be locally controlled and patterned on the material surface or throughout the bulk of the material. Also, whereas activating the material with stimuli such as heat can damage cells, the use of longer wavelengths of light, such as infrared and visible light, gives the advantage of activating the material with minimal and precicely controllable risk of harming attached cells. These dynamic, photo-activatable materials are the subjects of this chapter.

In this chapter, we will first give a brief overview of the types of materials being used to create photo-activatable cell biomechanics and mechanobiology platforms. Then, we will discuss ways in which photo-activatable materials have been applied to study cell attachment, migration, alignment and morphology, and differentiation. Finally, we will consider possible directions the field of photo-activatable materials for cell biomechanics and mechanobiology may take in the future.

4.2 Overview of Materials Used in Photo-Activatable Applications

The materials currently being developed as photo-activatable platforms for cell biomechanics or mechanobiolgy can, with a few exceptions, be grouped into three categories: self-assembled monolayers (SAMs), degradable hydrogels, and shape-memory polymers (SMPs). Here, the chemical groups typically added to these traditionally non-active materials in order to enable photoactivation will be discussed, and each material category will briefly be reviewed.

The approaches listed in the previous paragraph involve sensitizing a material to electromagnetic energy with the addition of a photo-responsive moiety. Typically, these moieties are either photo-cleavable and break a covalent bond upon irradiation, photoisomerizable and switch between isomer forms under irradiation, or photo-heatable and capable of absorbing electromagnetic energy and converting it

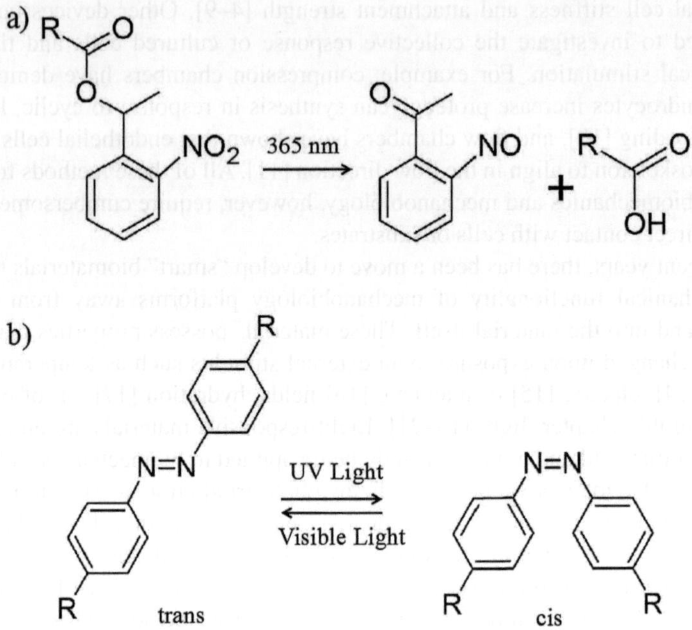

Fig. 4.2 (**a**) An example of nitrobenzyl photo-cleavage. In this case, the photo-cleavable group is nitrobenzyl ester. (**b**) Azobenzene isomerizes into its cis form upon UV irradiation and into its trans form under visible light irradiation

into heat (known as the photothermal effect). For example, derivatives of the nitrobenzyl group are the most commonly employed photo-cleavable group in the literature reviewed in this chapter (Fig. 4.2a). This group is capable of undergoing photolysis when exposed to 365 nm wavelength ultraviolet light, a wavelength that does not significantly reduce cell viability [22, 23]. The azobenzene group, a photoisomerizable group which switches between its cis and trans forms under UV and visible light irradiation, respectively, is also often added to sensitize materials to photo-irradiation (Fig. 4.1b). Finally, plasmonic nanoparticles and organic dyes are an example of photo-heatable elements that can be added to materials, causing them to heat when irradiated with the appropriate wavelength of light.

One class of materials highlighted in this chapter, self-assembled monolayers or SAMs, are materials with surfaces that have been chemically modified by the spontaneous adsorption of organic molecule chains. Typically, each chain is composed of an anchoring group that is strongly adsorbed onto the substrate surface (such as silicas on glass or a thiols on gold), followed by an alkyl chain capable of crystallizing with neighboring chains and stabilizing the SAM and a terminal functional group that gives the substrate desirable surface properties (e.g., hydrophobicity) [24]. SAMs have proven to be a useful tool in designing cell culture substrates due mainly to their ability to increase or decrease the substrate's ability to facilitate cell

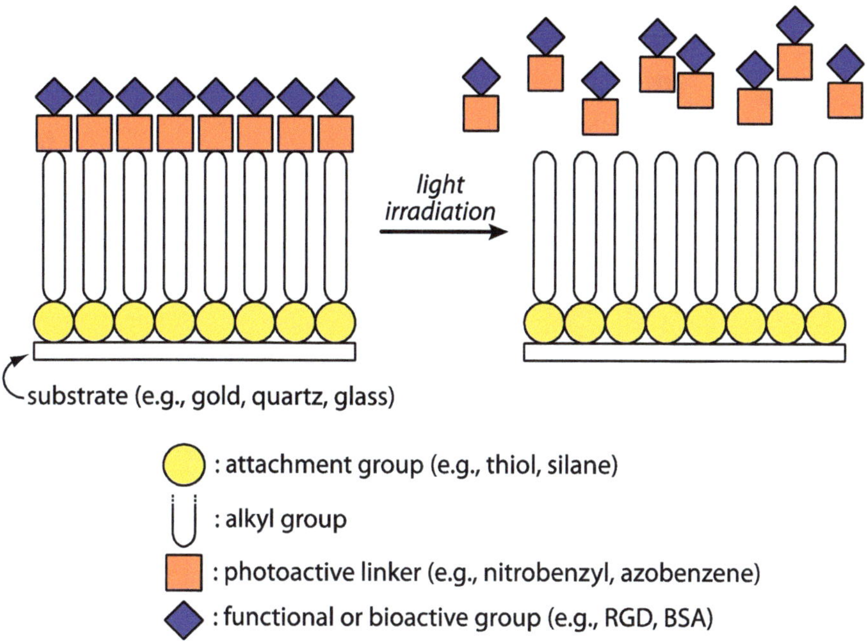

Fig. 4.3 A simplified depiction of a photo-activatable SAM. The SAM represented here is a simplified case where a photo-cleavable group (e.g., the nitrobenzyl group) links a cell-blocking (e.g., Bovine Serum Albumin (BSA)) or cell-adhesive (e.g., RGD peptide; see Sect. 4.3.1 for more detail) agent to the SAM surface. In this case, when irradiated with light, the linker and functional group cleave off, transforming the substrate surface from cell blocking to cell adhesive (if BSA is the functional group) or cell adhesive to cell blocking (if RGD is the functional group)

attachment [21] or present peptides to the cells that signal for the cells to alter their function [25].

The most common approach for preparing photo-activatable SAMs involves adding a photo-cleavable [21] or photoisomerizable [26, 27] group into the SAM's chains. When exposed to a certain wavelength of light, the photosensitive group cleaves or isomerizes, removing the original functional group and exposing a new functional group with different properties (Fig. 4.3).

Hydrogels are polymeric materials that swell in hydrophilic solutions as a result of having cross-linked polymer chains. These materials are commonly selected for biomaterials applications due to their biocompatibility, degradation, and mechanical properties similar to soft biological tissues [28]. Also, hydrogels can be cross-linked (liquid-gel transition) with cells present, allowing for the encapsulation of cells *inside* of hydrogel scaffolds. A clear advantage of encapsulation is that cells are present throughout the scaffold rather than simply on the surface, giving the scaffold three-dimensional features [29].

In order for a hydrogel to respond to photoactivation, it must include a photosensitive moiety in its polymer chains. Often, this moiety is a photo-cleavable [20, 30] or photoisomerizable [31] cross-linker. When photo-irradiated, the photodegradable

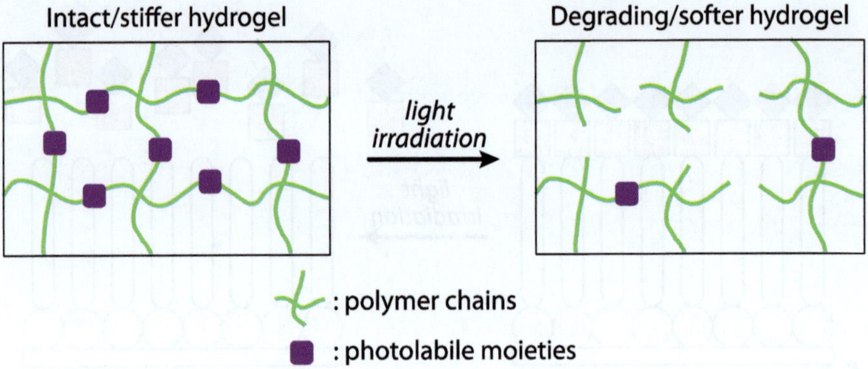

Fig. 4.4 A simplified depiction of a photodegradable hydrogel. This is the simplified case where a gel contains a single monomer with photo-cleavable cross-linker (e.g., a nitrobenzyl linker). Depending on the irradiation conditions, the gel can either be partially degraded resulting in a decrease in stiffness or fully degraded causing it to dissolve into solution

cross-links begin to break apart, degrading the hydrogel and reducing its stiffness (Fig. 4.4). If irradiated with substantially high intensity of light for substantially long periods of time (dependent on the material), the hydrogels can be fully degraded.

Shape-memory polymers (SMPs), unlike the other materials that have been discussed, are activatable materials by definition. SMPs are polymers capable of memorizing a permanent shape, being fixed in a temporary shape and then being triggered to recover back to their permanent shape when exposed to an external stimulus [32–35] such as heat [36–40], solvents [17, 41], electricity [42, 43], or light [19, 44–47]. The shape-memory mechanism involves a switch between mobile and fixed polymer chains and is often thermally driven, whether the applied stimulus is heat itself or an alternative stimulus that indirectly heats the polymer (e.g., light [19, 44–47] or electricity [42, 43]). Thermally triggered SMPs, for example, must have two characteristics: covalent or physical crosslinking that allow for "memorization" of a permanent shape (shape fixing) and a crystallization or vitrification event that allows for a switch in chain mobility at a specific melting (T_m) or glass transition (T_g) temperature (Fig. 4.5). The polymer is cross-linked into its permanent shape, heated above its transition temperature (either the T_m or T_g) to mobilize the polymer chains and allow for deformation into a temporary shape, and cooled below the transition temperature while holding the strain to fix the temporary shape. The polymer can then be triggered to recover back to its permanent shape when reheated above its transition temperature due to re-mobilizing the polymer chains and releasing the stored strain energy.

One difficulty associated with using SMPs in cell biomechanics and mechanobiology applications is that many cytocompatible SMPs, such as poly(caprolactone) and polystyrene, exhibit recovery temperatures well above physiological conditions and therefore cannot be triggered while maintaining cell viability [48]. More recently, however, we [38–40, 49–53] and others [46, 54–58] have developed SMPs

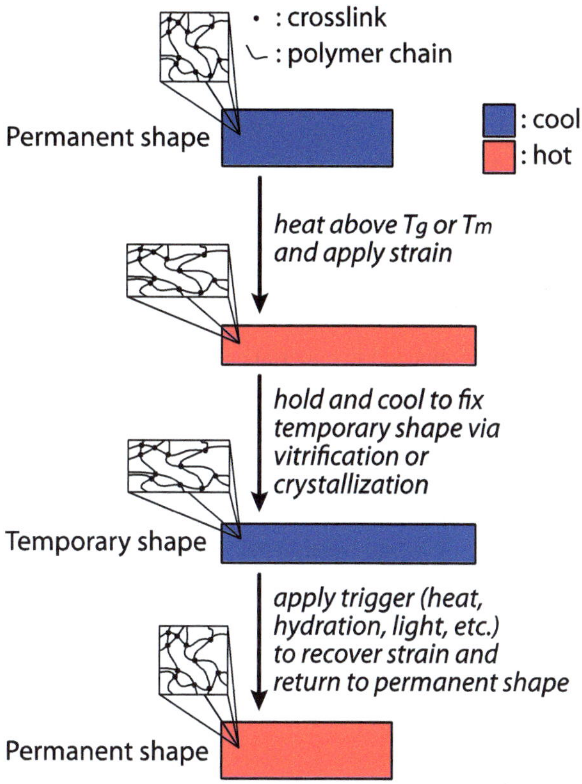

Fig. 4.5 Simplified depiction of the mechanism of thermally or photothermally activated SMPs. In this case, the polymer contains chemical cross-links that memorize the permanent shape and has a characteristic glass transition that allows molecule mobility above Tg and for switching between the temporary and permanent shapes. To sensitize this material to electromagnetic energy, dyes or nanoparticles are often added to increase the material's absorbance at a specific wavelength

capable of triggering under cell culture conditions and demonstrated that a change in substrate surface topography could be employed to control cell behaviors, such as alignment and motility [38, 49, 52–57, 59].

Photo-triggered shape-memory polymers are typically sensitized to light by adding either photo-cleavable cross-links to the polymer chain, as in the SMP developed by Lendlein and colleagues containing a cinnamic group [19], or by the introduction of a dye [44, 45] or plasmonic nanoparticles [46, 58] to the polymer that enhances its absorbance of light at a specific wavelength. In the latter method, which falls under the photothermal category of photoactivation, when the polymer is irradiated with the appropriate wavelength of light, it absorbs electromagnetic energy and converts it into heat, triggering SMP recovery. Photothermally activated SMPs have been applied to study cell biomechanics and mechanobiology and will be further discussed in the following sections [46].

The photo-activatable materials discussed in the remainder of the chapter fall into one of the three categories listed above: a photo-activatable SAM, photodegradable hydrogel, or photothermally activated SMP. All three of these material types have been implemented in cell biomechanics or mechanobiology in order to study how altering the substrate's surface chemistry, topography, or stiffness can influence cell attachment (Sect. 4.3.1), cell migration (Sect. 4.3.2), cell alignment and morphology (Sect. 4.3.3), or differentiation (Sect. 4.3.4).

4.3 Applications of Photo-Activatable Materials to Study Cell Biomechanics and Mechanobiology: Influencing Cell Attachment, Migration, Morphology/Alignment, and Differentiation

Having introduced the material categories commonly used in photo-activatable cell biomechanics and mechanobiology studies, we now focus on ways in which these materials have been applied to direct and study cell behavior. The discussion is organized in four subsections highlighting different cell behaviors that have been studied: cell attachment, cell migration, cell alignment and morphology, and cell differentiation. It is important to note, however, that many of the studies discussed investigated not one but multiple cell behaviors as well as behaviors that may not fall neatly into one of the aforementioned categories. Also, the field of photo-activatable cell biomechanics and mechanobiology is still in its infancy, and much of the work to date has focused on materials and concept development and demonstration. The full potential of these materials is still largely unrealized, and possible avenues in which the field may travel in the future are discussed in the final section of this chapter.

4.3.1 Photo-Activatable Materials for Influencing Cell Attachment

Cell attachment plays a critical role in a cell's interaction and communication with its environment. In cell biomechanics and mechanobiology specifically, cell attachment sites are important because they facilitate the exchange of mechanical forces between a cell and its substrate in vitro. A number of researchers interested in cell biomechanics and mechanobiology have developed materials whose surfaces can be actively altered during cell culture using photo-irradiation, and the results of these studies are discussed in this section.

Manipulation of cell attachment using photo-irradiation has been widely employed in cell patterning [21, 25, 60–70]. Nakanishi and colleagues [21] were one of the earliest groups to pattern cells using photo-controlled modification of cell attachment sites. They demonstrated the ability to pattern down to the precision of a single cell by irradiating cell-sized (25 μm) sections of the substrate and to actively produce new cell adhesion locations during cell cultivation (Fig. 4.6). In order to do this, the group implemented a SAM approach by adsorbing a molecule containing 2-nitrobenzyl ester and capped with bovine serum albumin (BSA) (Pluronic F-108

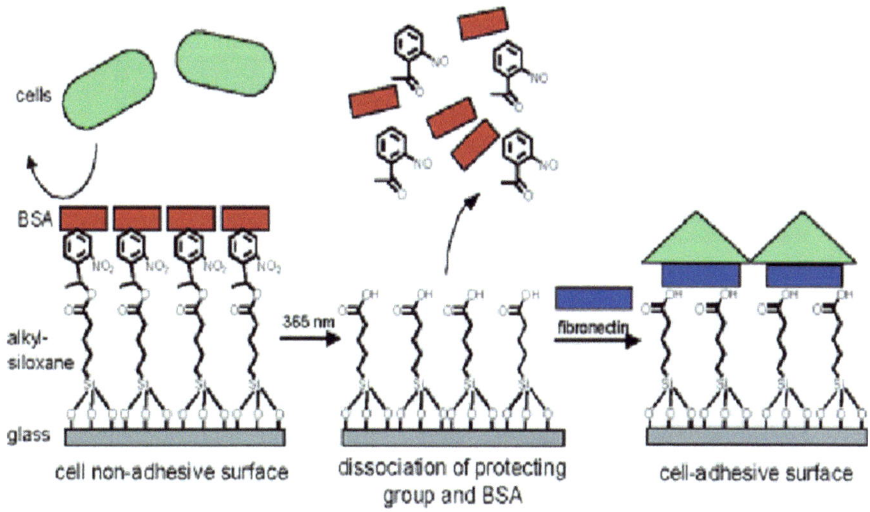

Fig. 4.6 Process behind photo-activatable SAM used by Nakanishi [21]. Initially, BSA is adsorbed to the surface and cells cannot adhere. Then, the nitrobenzyl group cleaves in UV-irradiated regions of the substrate surface, removing the BSA. Fibronectin is then adsorbed in the irradiated regions, and cells are able to adhere to the surface (*Source*: Adapted and reprinted with permission from (Nakanishi et al., Photoactivation of a substrate for cell adhesion under standard fluorescence microscopes. *J. Am. Chem. Soc.* 126, 16,314–16,315 (2004)). Copyright (2004) American Chemical Society)

has also been adsorbed onto the surface as the attachment blocking agent [62, 69]), to the surface of a glass substrate. When irradiated with 365 nm light, the nitrobenzyl group is cleaved, removing the BSA and exposing a more hydrophilic carboxyl group. Then, fibronectin, a protein known to improve cell attachment, is adsorbed to the surface. With multiple irradiations, a single cell can be adhered in proximity to another previously attached cell. Since the adhesion sites on this material can be modified during cell culture, this platform has proven useful in dynamic cell patterning applications and in patterning multiple cell types together (Fig. 4.7). This same method has also been used to create smaller, subcellular-sized regions (~6 μm) of cell attachment on substrates in order to pattern cell adhesion points (Fig. 4.8) [61].

One disadvantage to using protein adsorption as an attachment blocking/facilitating agent is that the patterning may fade over time as the adsorbed proteins exchange with others present in the cell culture media. For example, if the cell attachment blocking proteins (e.g., BSA) are exchanged with non-blocking proteins, cells begin to migrate out of confined pattern regions over time. In order to improve pattern longevity and remove the need for BSA or Pluronic adsorption, photo-cleavable blocking agents such as poly(ethylene glycol) (PEG) have been covalently attached to the SAM chains using a nitrobenzyl group [25, 63, 64, 66, 67]. Following UV irradiation, these blocking agents are cleaved off and allow for cell attachment. Kikuchi and colleagues found that switching from protein

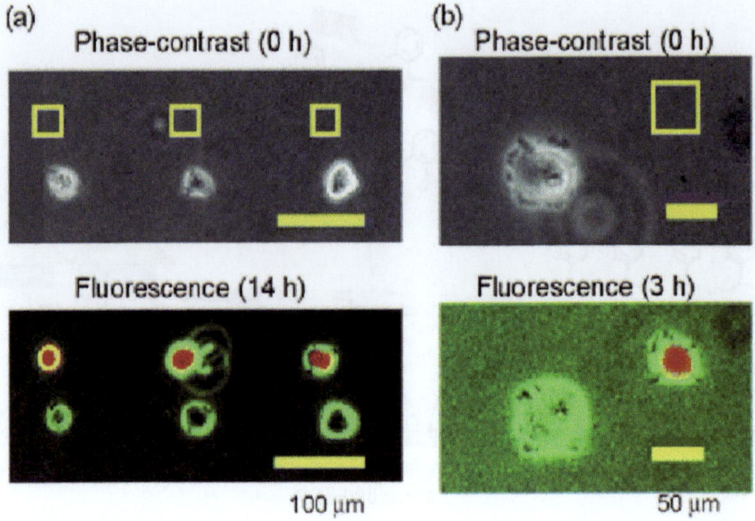

Fig. 4.7 Phase-contrast and fluorescence images of cells patterned using a photo-activatable substrate. (**a**) HEK293 cells were attached in advance using the first UV irradiation, and then, fluorescent HEK293 cells were attached in the highlighted boxes (phase-contrast image) with a subsequent irradiation. (**b**) A COS7 cell was initially attached to the surface, and then a fluorescent HEK293 cell was attached to the area highlighted with a box (phase-contrast image) with a subsequent irradiation (*Source*: Adapted and reprinted with permission from Nakanishi et al., Photoactivation of a substrate for cell adhesion under standard fluorescence microscopes. *J. Am. Chem. Soc.* 126, 16,314–16,315 (2004). Copyright (2004) American Chemical Society)

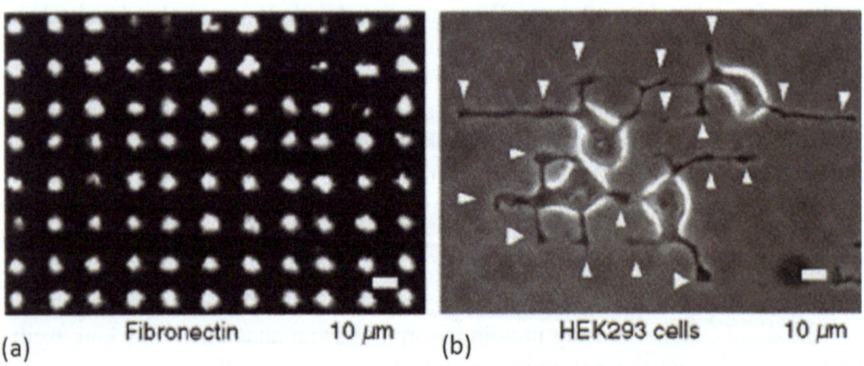

Fig. 4.8 HEK293 cell patterning on photo-activated subcellular-sized attachment sites. (**a**) Immunofluorescence image of photo-patterned fibronectin attachment sites. Attachment sites are approximately 6 µm by 6 µm. (**b**) A phase-contrast image of HEK293 cells seeded onto the photo-patterned substrate with nodal structures (represented by arrowheads) attached at the irradiated regions (*Source*: Adapted and reprinted from *Analytica Chimica Acta*, 578, Nakanishi et al., Spatiotemporal control of cell adhesion on a self-assembled monolayer having a photocleavable protecting group, 100–104, Copyright (2006), with permission from Elsevier)

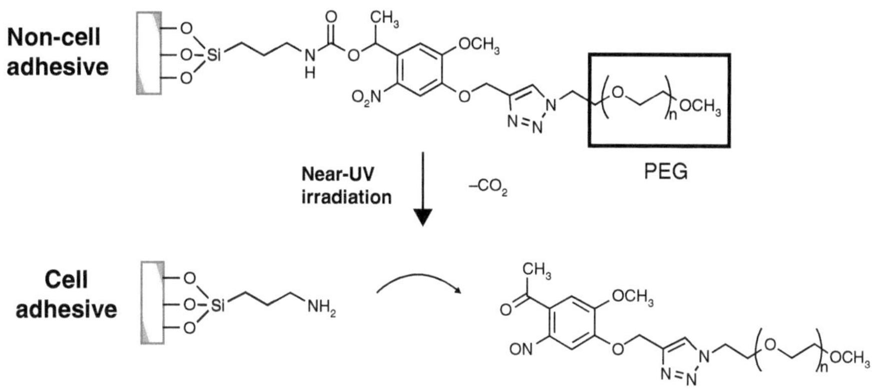

Fig. 4.9 An example of a photo-activatable SAM that does not require protein adsorption. The photo-cleavable PEG group blocks cell attachment. However, when irradiated with UV light, the PEG cleaves off, exposing the -NH_2 group and allowing for cell attachment (*Source*: Adapted and reprinted from Kaneko, S. (2011) with permission of the Royal Society of Chemistry)

adsorption to covalently linked PEG improved cell pattern longevity from 3 to 17 days [64]. Kikuchi and colleagues also explored an alternative to fibronectin adsorption to facilitate cell attachment by replacing the caged carboxylic acid (COOH) group with an amine group (NH_2) [65]. The amino group allows for cell attachment without any subsequent protein adsorption. Also, the photo-cleavable PEG and amino cell attachment methods have been combined so that the surfaces can switch from cell blocking to cell adhesive without any protein adsorption requirements (Fig. 4.9) [66].

As an alternative to full proteins, small protein fragments, or peptides, have been incorporated into photo-activatable SAM platforms to facilitate cell attachment [60, 68, 70, 71]. The RGD peptide, which contains the arginine-glycine-aspartate amino acid sequence, is the portion of fibronectin and other ECM proteins responsible for cell binding, and this sequence is independently capable of binding to cells [72]. Both Ohmuro-Matsuyama [71] and Peterson [68] have used RGD peptides as the cell-binding portion of their photo-activatable SAM platforms in order to actively pattern cells. In both cases, the RGD peptide was caged with a photo-cleavable nitrobenzyl group to prevent cell adhesion. When irradiated with UV light, the nitrobenzyl cage is removed, exposing the RGD sequence and allowing cells to bind selectively to irradiated sections. Han and colleagues [25] developed a SAM containing the PFSSTKTC (Pro-Phe-Ser-Ser-Thr-Lys-Thr-Cys) peptide, which is capable of binding cells as well as mediating stem cell differentiation [73, 74]. The peptide is blocked by an o-nitrobenzyl containing PEG group, which sterically hinders cell attachment. However, when exposed to 365 nm light, the nitrobenzyl group is photo-cleaved and peptide is exposed, allowing for cell attachment. The authors demonstrated that they could confine cell attachment to irradiated regions on the substrate surface and then further investigated their ability to confine the differentia-

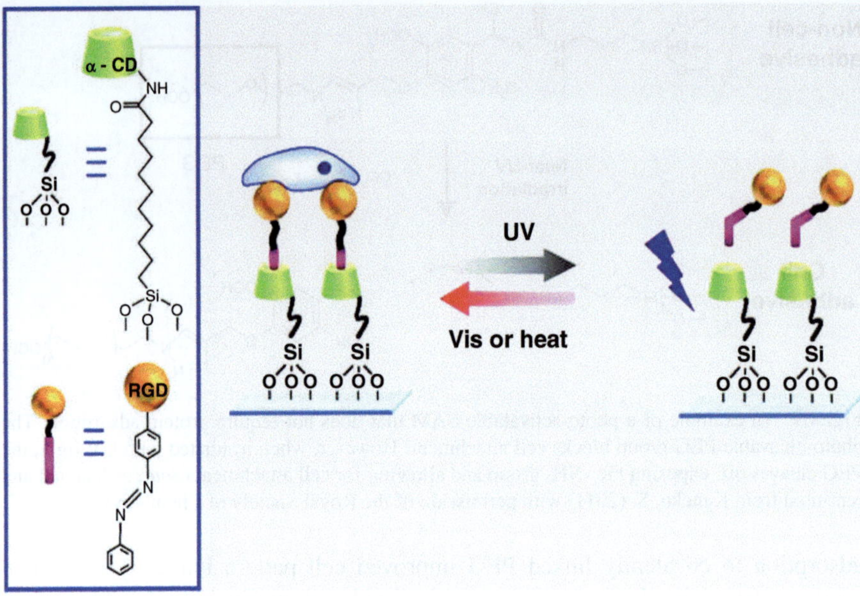

Fig. 4.10 An example of an azobenzene-containing SAM that allows for reversible cell attachment depending on lighting condition. Under visible light, the azobenzene molecule is in its trans (E) form and can bind to the α-cyclodextrin (α-CD) moiety, therefore attaching to the SAM and allowing for cell attachment on RGD sites on the SAM surface. However, under UV irradiation (365 nm), the azobenzene isomerizes into its cis (Z) form and detaches from the SAM, preventing cells from attaching to the surface. Returning to visible light reverts the azobenzene back to its trans form, again allowing for cell attachment (*Source*: Adapted and reprinted with permission from Gong, Y.H. et al. Photoresponsive 'Smart Template' via Host-Guest Interaction for Reversible Cell Adhesion. *Macromolecules* 44, 7499–7502 (2011)). Copyright (2011) American Chemical Society)

tion of mesenchymal stem cells (MSCs) to specific regions on the substrate, and this will be discussed in more detail in the cell differentiation section of this chapter.

The photo-activatable cell attachment platforms discussed so far are limited in that, once cell attachment sites are created by photo-irradiation, they cannot be removed. To address this, *reversible* photo-activatable platforms for studying cell attachment have been developed that implement the photoisomerizable azobenzene moiety. Cell attachment can be selectively turned off in specific locations and then turned back on by changing the light irradiation conditions between UV and visible wavelengths (Fig. 4.10) [26, 27]. In order to achieve this functionality, SAMs containing azobenzene groups linked to an RGD adhesion peptide (arginine-glycine-aspartate) are adhered to the substrate (either quartz [26] or gold [27]). The azobenzene molecule switches from its trans form to its cis form upon exposure to UV irradiation, either detaching [26] or caging [27] the RGD peptide so that it can no longer make contact with cells. Then, when subsequently irradiated with visible light, the molecule reverts back into its trans form and once again allows cells to adhere to the surface.

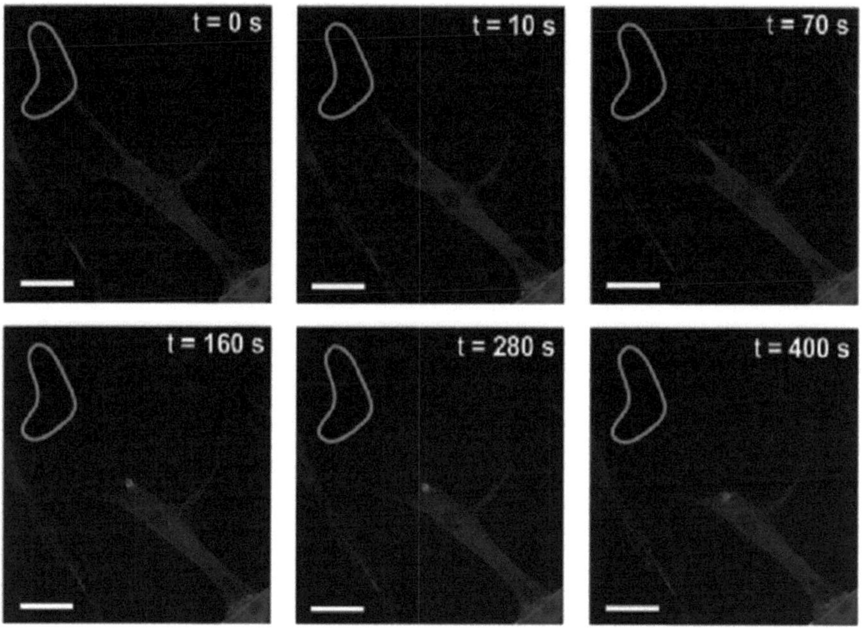

Fig. 4.11 A specific cell attachment region (t = 0 s, region circled in gray) of the hydrogel was eroded with two-photon irradiation. Decrease in cell area in response to attachment site removal was tracked for up to 400 s postirradiation. Scale bar represents 20 μm (*Source*: Reproduced in part from Tibbett et al. (2010) with permission of the Royal Society of Chemistry)

All of the methods for photo-modification of cell attachment discussed so far have involved photo-activatable SAMS; however, the photodegradation of hydrogel substrates has also been implemented to study how cells retract after the removal of attachment sites (Fig. 4.11) [75]. In order to accomplish this, Tibbett and colleague seeded MSCs onto photodegradable hydrogels containing photo-cleavable cross-linkers containing nitrobenzyl ether. Using two-photon irradiation, 740 nm light was focused down to subcellular (micron-scale) size and targeted directly beneath cell attachment sites, causing the gel to degrade and the cell to retract. By monitoring the cell area postirradiation, the authors were able to determine that MSCs detached six times slower on the soft hydrogels than on much stiffer glass substrates [75, 76].

In most of the studies discussed, photo-activatable materials were used to demonstrate the ability to actively pattern cells onto a substrate. The use of photo-irradiation as an activation stimulus allowed for remotely and precisely controlling the placement of cells on a substrate and also allowed for changing the cell placement in time with subsequent irradiations. Cell patterning is a useful tool on its own, for example, to select where different cell types are attached on a tissue-engineered implant or to study how different cell types interact with one another in various patterns in vitro. However, in many studies that will be discussed in the following sec-

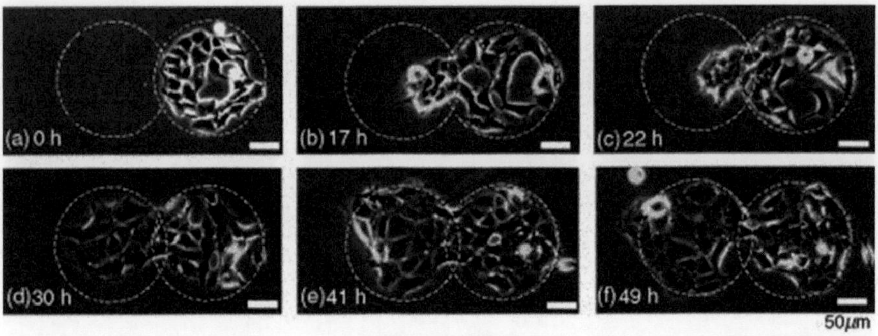

Fig. 4.12 An example of inducing cell migration by photo-creation of new cell attachment sites during cell culture. Initially, HEK293 cells attached at the circular spot on the right with the first photo-irradiation. Then, a second circular region (left) was illuminated with UV light to induce cell migration toward the left circular region (*Source*: Adapted and reprinted from *Analytica Chimica Acta*, 578, Nakanishi et al., Spatiotemporal control of cell adhesion on a self-assembled monolayer having a photocleavable protecting group, 100–104, Copyright (2006), with permission from Elsevier)

tions, photo-activatable cell attachment platforms were further applied to control other cell activities such as migration, morphology, and differentiation.

4.3.2 Photo-Activatable Materials for Influencing Cell Migration

A cell's ability to migrate through ECM is important in many biological processes, such as embryonic development and wound repair [77]. Also, changes in migration have been associated with cancer metastasis [78]. Therefore, many platforms have been developed to study and direct cell migration. Two methods that have allowed researchers to study cell migration involved either actively creating new cell attachment sites on the substrate surface (Fig. 4.12) [61, 62, 64, 67, 70] or degrading migration channels into a cell-encapsulated hydrogel [20]. A number of studies have focused on developing substrates capable of altering cell migration behavior in response to electromagnetic irradiation [20, 61, 62, 64, 67, 70]. Highlights from these studies will be discussed in this section.

Photo-activatable substrates have been used to induce cell migration along different sized paths and to investigate the effect of path size on migration rate [62]. In order to accomplish this, a platform was developed by Naknishi and colleagues that implemented photo-selective detachment of BSA and subsequent adsorption of fibronectin on the substrate surface [21]. First, individual cells were photo-patterned onto the surface by irradiating 25 μm squares with UV light. Then, adjacent regions were irradiated to create either 5 μm or 25 μm wide channels of cell-adhesive surface. This allowed the cells to migrate along channels of different sizes. It was

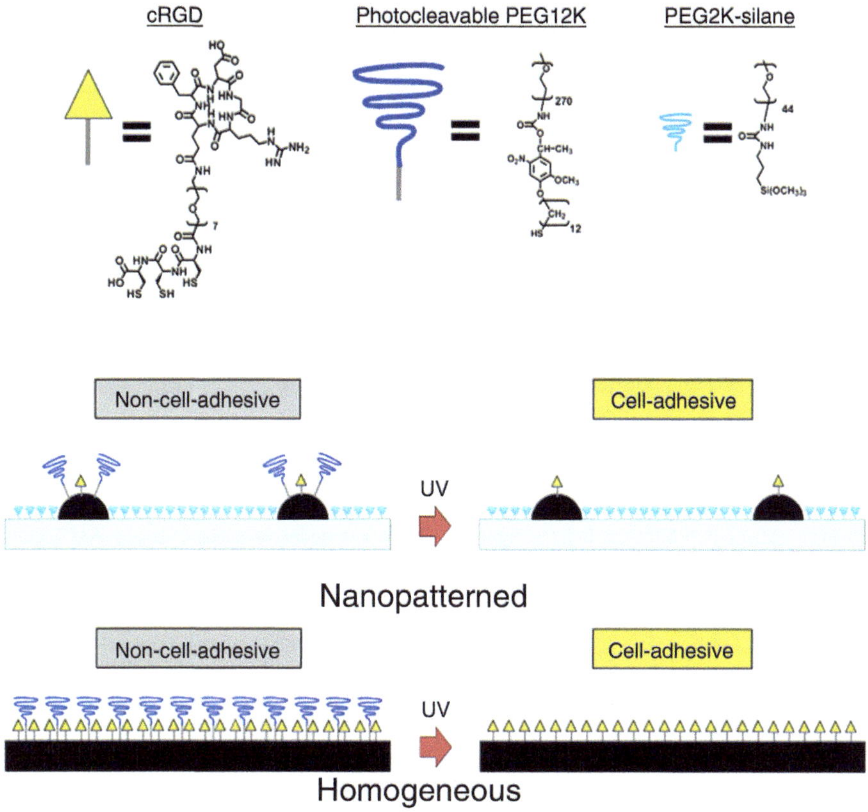

Fig. 4.13 For both the nanopatterned and homogeneous substrates, the RGD is buried beneath photo-cleavable 12 K molecular weight PEG, preventing cell attachment. However, when exposed to UV irradiation, the PEG is removed and allows cells to make contact and attach to the RGD. For the nanopatterned substrate, cells can only attach in a pattern predetermined by the placement of the gold nanoparticles (half circles), but on the homogeneous substrate, cells can attach at any irradiated region of the surface (*Source*: adapted and reprinted with permission from Shimizu, Y., Boehm, H., Yamaguchi, K., Spatz, J. P., & Nakanishi, J. A photoactivatable nanopatterned substrate for analyzing collective cell migration with precisely tuned cell-extracellular matrix ligand interactions. *PLOS One* 9(3) e91875, https://doi.org/10.1371/journal.pone.0091875 (2014))

determined that NEK3T3 fibroblast cells migrated faster on wider (25 μm) paths than thinner (5 μm) paths most likely due to the fact that the thinner path obstructed the cells' cytoskeletal elongation.

Photo-activatable SAM platforms have also allowed for researchers to study the collective migration of HeLa cells. Shimizu and colleagues [70] developed a nanopatterned SAM for studying cell migration by arranging gold nanoparticles at specific locations on a substrate surface (Fig. 4.13). Then, RGD peptides were selectively attached to the gold nanoparticles. Nitrobenzyl ester-terminated PEG was also attached to the gold nanoparticles, forming a cage around the RGD peptide and preventing cells from attaching to the surface. However, when exposed to UV

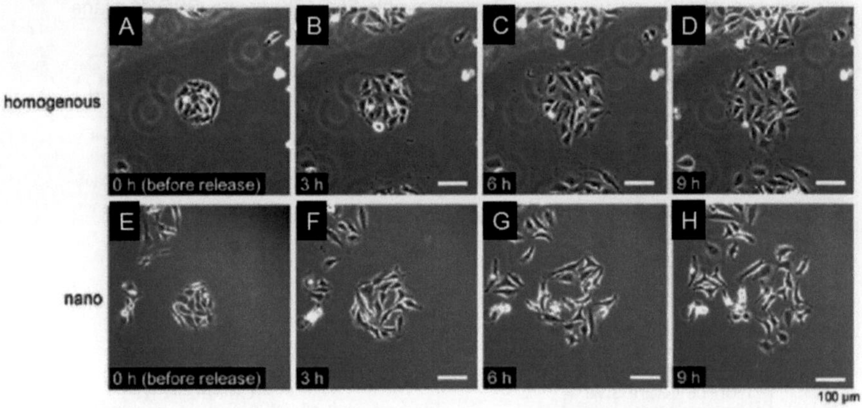

Fig. 4.14 The migration of HeLa cells was observed using both the nanopatterned and homogeneous substrates depicted in Fig. 4.13. Initially, a 150 µm circular region of attachment was irradiated onto the surface, and cells were allowed to adhere. Then, the entire surface was irradiated to form new attachment sites and induce migration. On the homogeneous surface (images A–D), cells migrated out radially but maintained cell–cell interactions. On the nanopatterned surface (images E–H), cells migrated away from one another and lost their collective migration tendency (*Source*: adapted and reprinted with permission from Shimizu, Y., Boehm, H., Yamaguchi, K., Spatz, J. P., & Nakanishi, J. A photoactivatable nanopatterned substrate for analyzing collective cell migration with precisely tuned cell-extracellular matrix ligand interactions. *PLOS One* 9(3) e91875, https://doi.org/10.1371/journal.pone.0091875 (2014))

light, the PEG cleaved off the surface and allowed cells to attach in specific locations predetermined by the gold nanoparticle pattern. Cell migration was induced by actively creating new attachment sites during cell culture by exposing certain areas of the surface to UV irradiation. The cell locations were manually tracked in time under microscope observation, and then the location data was used to calculate migration rates. The authors demonstrated that HeLa cells lose their collective migration tendencies on surfaces with fewer, patterned attachment sites (see "nanopatterned" substrate in Fig. 4.13) compared to surfaces with a higher number of randomly arranged attachment sites (see "homogeneous" substrate in Figs. 4.13, 4.14). This finding contradicts the conventional theory that epithelial (including HeLa) cells migrate more collectivley on surfaces with lower cell adhesiveness [79].

Photo-activatable materials have proven useful in studying the effect of different drugs on cell migration [67]. For example, researchers developed a photo-activatable cell culture platform that could be used as a high-throughput platform for screening cell migration inhibiting or accelerating drugs. In this case, the bottom of the glass well plate was coated with a SAM containing nitrobenzyl ester-termintated PEG groups. Similar to the previously described study, new cell attachment sites could be created during cell culture with UV irradiation to facilitate cell migration. Using this new substrate, the authors were able to demonstrate that cytochalasin D, a cell migration inhibitor, is capable of inhibiting cell cluster expansion and further

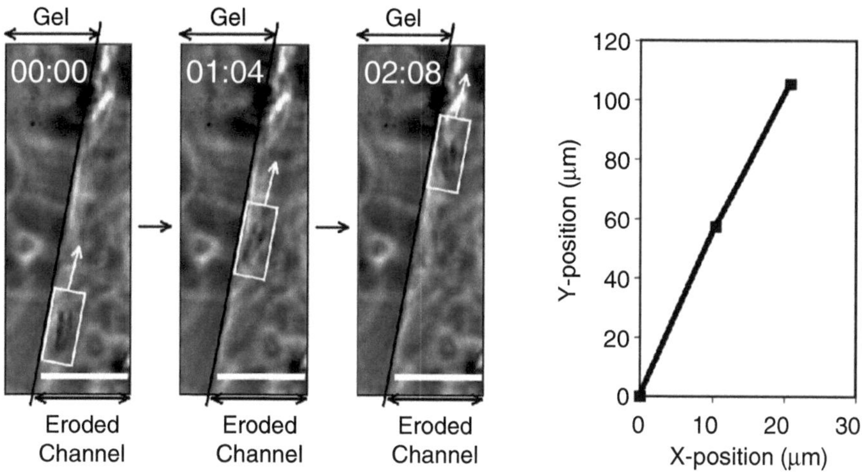

Fig. 4.15 Migration channels were created inside of a hydrogel filled with encapsulated fibrosarcoma cells by degrading the gel with photo-irradiation. Bright-field images of a cell (highlighted in the white box) migrating along the edge of a channel are shown on the left. The cell's position in time is plotted on the right. Scale bar represents 50 μm (*Source*: From Kloxin, A. M., Kasko, A. M., Salinas, C. N., & Anseth, K. S. Photodegradable Hydrogels for Dynamic Tuning of Physical and Chemical Properties. *Science*. 324, 59–63 (2009)). Reprinted with permission from AAAS)

showed that cells cluster expansion rates either increased or decreased in response to blebbistatin, a myosin II inhibitor, depending on the cluster width.

The photo-activatable cell migration platforms discussed up to this point have all been two-dimensional in nature; however, cells in their natural environment migrate in three dimensions. Therefore, materials that allow for studying three-dimensional migration of cells in vitro have been developed using the light-induced degradation of hydrogels. Kloxin and colleagues [20] encapsulated fibrosarcoma cells in their PEG hydrogel containing photodegradable cross-links. Using UV light, the gel could be irradiated at any 3D location, causing it to completely degrade and allowing for the cells to migrate along the newly formed channels (Fig. 4.15). The group demonstrated that their platform could be used to direct cell migration in real time.

4.3.3 *Photo-Activatable Materials for Influencing Cell Morphology and Alignment*

In living tissues, cells are often aligned (oriented) in specific directions with distinct morphologies that vary from tissue to tissue. For example, endothelial cells in blood vessels align in the direction of blood flow [11], and skeletal muscle cells align along the same axis to form a contracting muscle [80]. Therefore, scientists have attempted to direct cell morphology and orientation in vitro to better study and

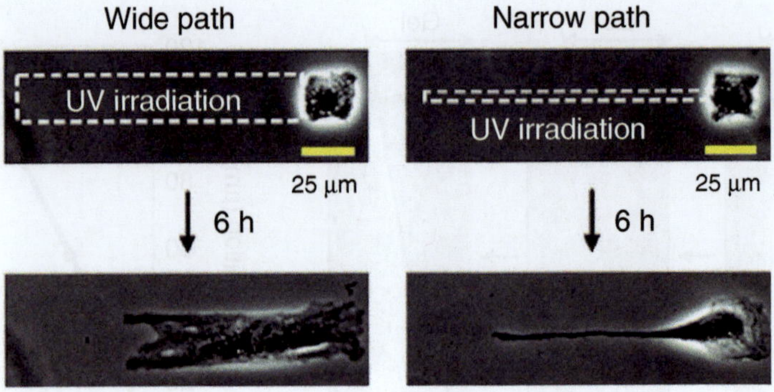

Fig. 4.16 Morphology of NIH3T3 fibroblast cells was modified via photoactivation of attachment sites. Cells were initially attached to the substrate by irradiating a 25 μm by 25 μm region of the substrate. Then, with a subsequent irradiation, either a 25 μm wide (left, wide path) or 5 μm wide (right, narrow path) region of attachment was created. The UV irradiation regions are outlined as dotted rectangles. Flat protrusions, lamellipodia, were formed along the 25 μm channel, while thin protrusions, filopodia, were formed along the 5 μm channel (*Source*: Adapted and reprinted with permission from Nakanishi et al., Spatiotemporal control of migration of single cells on a photoactivatable cell microarray. *J. Am. Chem.* Soc. 129, 6694–6695 (2007)). Copyright (2007) American Chemical Society)

mimic living tissues. The use of photo-activatable materials to direct cell morphology and alignment will be discussed in this section.

Photo-activatable SAMs similar to those described in previous sections have been used to actively alter cell morphology by controlling cell attachment sites. Nakanishi and colleagues confined NIH3T3 fibroblast cells into a triangular morphology by irradiating their SAM substrate (See Sec. 4.3.1) with UV light in order to create cell-sized, triangular adhesive regions [69]. Once cells were confined, cell protrusions were monitored. It was discovered that the protrusions extended preferentially from the sharp edge of the triangle. This same platform was used to study how cell protrusions extend along micron-sized channels (Fig. 4.16) [62]. Initially, NIH3T3 fibroblast cells were patterned onto 25 μm squares by creating cell adhesion regions by UV irradiation. Then, a subsequent irradiation was used to create a small (5 μm) or large (25 μm) adhesive region adjacent to previously attached cells. The authors found that cells extended thin protrusions, called filopodia, into the smaller channels. However, they extended flat protrusions, called lamellipodia, into the larger channel and spread over the entire cell adhesive region. It was also demonstrated that the actin stress fibers (components of the cell cytoskeleton) were aligned parallel to the thin channel but randomly aligned in the wider channel. Dynamically changing stress fiber alignment could be used in the future to direct cell migration, since the stress fiber alignment is associated with a cell's forward motion.

Photodegradable hydrogels have also been used to actively alter cell morphology. Frey and Wang [30] developed a polyacrylamide hydrogel containing photocleavable 4-bromomethyl-3-nitrobenzoic acid (BMBA) cross-links in order to study how substrate stiffness affects cell morphology. The BMBA, which contains a nitrobenzyl group, undergoes photolysis under 365 nm light irradiation, breaking down cross-links and softening the gel. The authors seeded NIH3T3 fibroblasts onto the surface of the hydrogel and performed both total and partial cell irradiation. The irradiation causes the gel to partially degrade and soften underneath the cell. It was determined that cell spread area decreased under full-cell irradiation. However, when only part of the cell was irradiated, cell area decreased when the anterior region of the cell was irradiated but exhibited less decrease with posterior irradiation (Fig. 4.17). They concluded that mechanosensing of NIH3T3 fibroblasts may be focused in anterior section of cells. Kloxin and colleagues also investigated the effects of stiffness on cell morphology using a photodegradable hydrogel [22]. However, in this case, MSCs were encapsulated into the gel so that 3D morphologies could be observed. A stiffness gradient was created in one direction by controlling the intensity of light irradiating different portions of the gel. Cells were able to take on a more spread-out morphology in the more highly degraded regions of the gel, and cell spread area correlated with the degradation gradient.

Changes in cell alignment have also been directed with photo-activated changes in a materials' surface topography. For example, UV light was used to degrade ridges (5 μm wide and 1 μm deep) into the surface of a photodegradable hydrogel with valvular interstitial cells (VICs) seeded on the surface [81]. The change in topography caused the VIC cells to align parallel with the ridges and also caused cells on soft gels (modulus~3kPA) to change their phenotype into myofibroblasts. This change in phonotype will be discussed further in the following section (see Sect. 4.3.4). Photosensitive shape-memory polymers have also been a useful tool in actively changing cell alignment [59]. In this case, gold nanorods, which absorb NIR light, were incorporated into a poly(caprolactone) (PCL) SMP so that the polymer would undergo a shape change upon the absorption of NIR light. SMP was fixed into a temporary ridged shape with nanoscale grooves, and NIH3T3 fibroblasts were seeded onto the surface. At first, cells aligned with the grooves; however, upon partial exposure to NIR irradiation, the SMP recovered its permanent (flat) shape, and cell alignment decreased locally in the area of irradiation.

4.3.4 *Photo-Activatable Materials for Influencing Cell Differentiation and Phenotype*

Stem cells are capable of sensing mechanical cues from their environment and using these cues to direct their differentiation fate and final phenotype [82]. Therefore, materials capable of actively changing their mechanical properties (e.g., stiffness) or peptide presentation have recently been developed that allow researchers to study how these changes affect stem cell fate. In this section, photo-activatable platforms for studying cell differentiation and phenotype changes will be discussed.

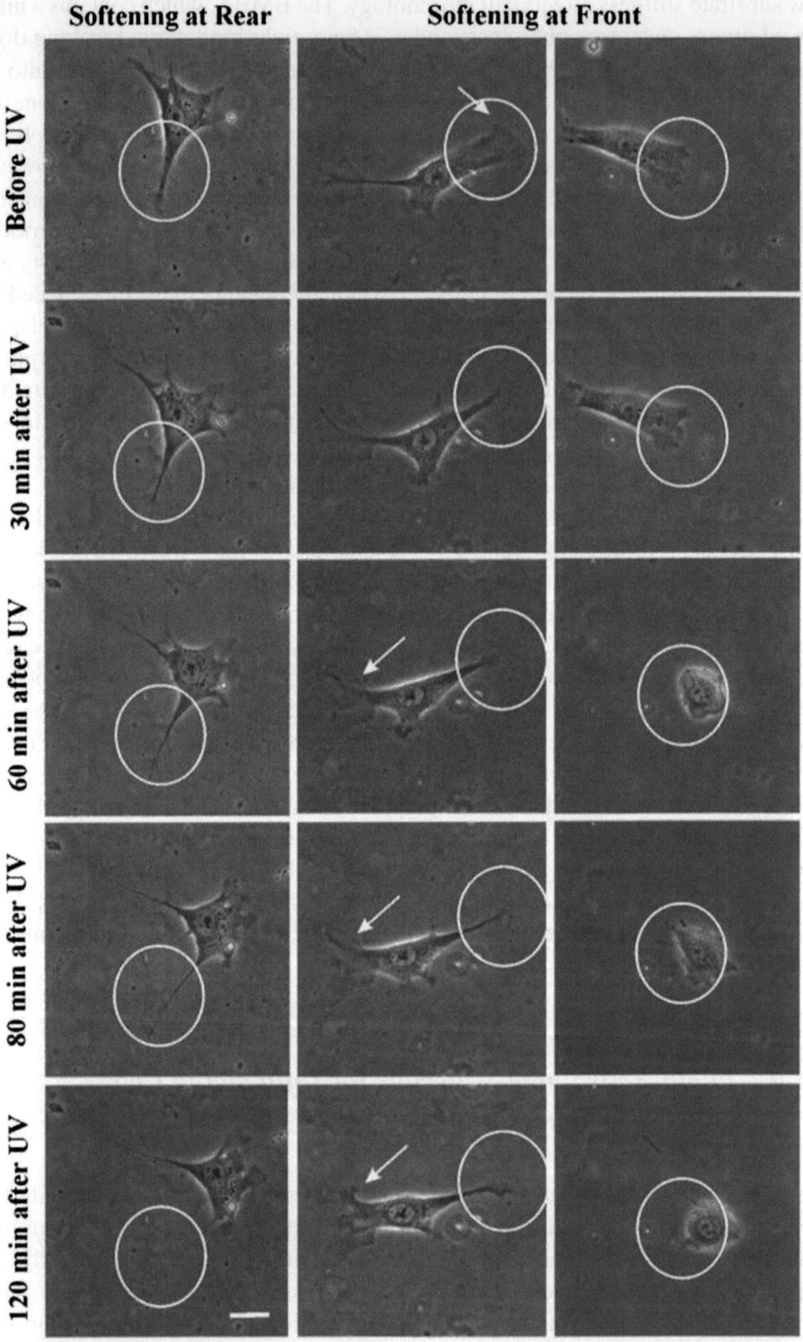

Fig. 4.17 Images showing NIH3T3 cell response to photo-activated decrease in hydrogel stiffness localized at either the posterior (rear, left column) or anterior (front, center, and right columns) of the cell. Decreasing the stiffness at the posterior of the cells did not affect cell spread area (left column). However, decreasing the stiffness at the anterior of the cell caused cells to reverse polarity (middle column) or become trapped in the softened area of the gel (right column). Scale bar represents 20 μm and the irradiated region shown as white circles (*Source*: Reproduced from Frey and Wang (2009) with permission from the Royal Society of Chemistry)

One method that has been used to induce stem cell differentiation is the photo-controlled presentation of certain peptides to stem cells. Han and colleagues [25] developed a photo-activatable SAM capable of attaching and differentiating MSCs in specific locations on the substrate surface by presenting the PFSSTKTC peptide (see Sect. 4.3.1) upon exposure to UV irradiation. As previously discussed in the cell attachment section, this peptide has been shown to assist in differentiating MSCs into osteoblasts (bone cells) [73, 74]. The authors demonstrated that they could spatially control regions of MSC to osteoblast differentiation on their substrate surface. Control over differentiation using peptide exposure has also been achieved using photodegradable hydrogels [20]. In this situation, an RGDS (Arg-Gly-Asp-Ser) peptide was covalently attached to the hydrogel via a photo-cleavable linker. When irradiated with UV light, the linker breaks and releases the RGDS. RGDS concentration can be controlled in both space and time by altering the irradiation conditions. It is important to note that, although fibronectin, which contains the RGDS peptide, is most often associated with facilitating cell attachment, it also plays a role in the differentiation of MSCs into chondrocytes (cartilage cells) [83]. MSCs secrete fibronectin early on during their differentiation into chondrocytes. MSCs were encapsulated in the photodegradable hydrogel, and, by controlling the release of RGDS by irradiation, the cells were successfully differentiated into osteoblasts.

Hydrogels capable of undergoing changes in stiffness under photo-irradiation have allowed researchers to study how changes in stiffness affect the activation of valvular interstitial cells (VICs) into myofibroblasts [31, 81, 84, 85]. Before discussing the details of these studies, it is first necessary to explain the natural function of VICs. VICs are a type of fibroblast that reside in the aortic valve and are responsible for connective tissue maintenance and repair. When tissue damage occurs, VICs adopt an activated, myofibroblast phenotype. The myofibroblast phenotype excretes ECM proteins and develops a contractile cytoskeleton in order to repair damaged tissue. In healthy tissue, once repair is complete, the myofibroblasts deactivate back into the quiescent VIC state. However, diseases develop when the myofibroblast phenotype persists and causes excessive tissue stiffening. Therefore, an understanding of the transition between the two VIC phenotypes can give insight into these types of diseases. In order to study VIC activation, Kloxin and colleagues seeded VICs onto initially stiff (Young's Modulus $E \sim 30$ kPa) photodegradable hydrogels. The stiff nature of the surface caused cells to activate into the myofibroblast phenotype. Upon UV irradiation, the gel's stiffness decreased ($E \sim 7$ kPa) and cells reverted back to the VIC phenotype, thus demonstrating that myofibroblasts can be dedifferentiated back into VICs solely by decreasing the substrate stiffness [84] (Fig. 4.18). It was further demonstrated that deactivated VICs on the soft substrates could be partially reactivated either by administering differentiation- and proliferation-inducing growth factors [85] or forming micron-sized grooves on the gel's surface by patterned photodegradation [81].

Hydrogels that undergo *reversible* softening in response to UV irradiation have also been used to investigate VIC activation in response to substrate stiffness changes [31]. In contrast to other hydrogels utilized in VIC differentiation studies

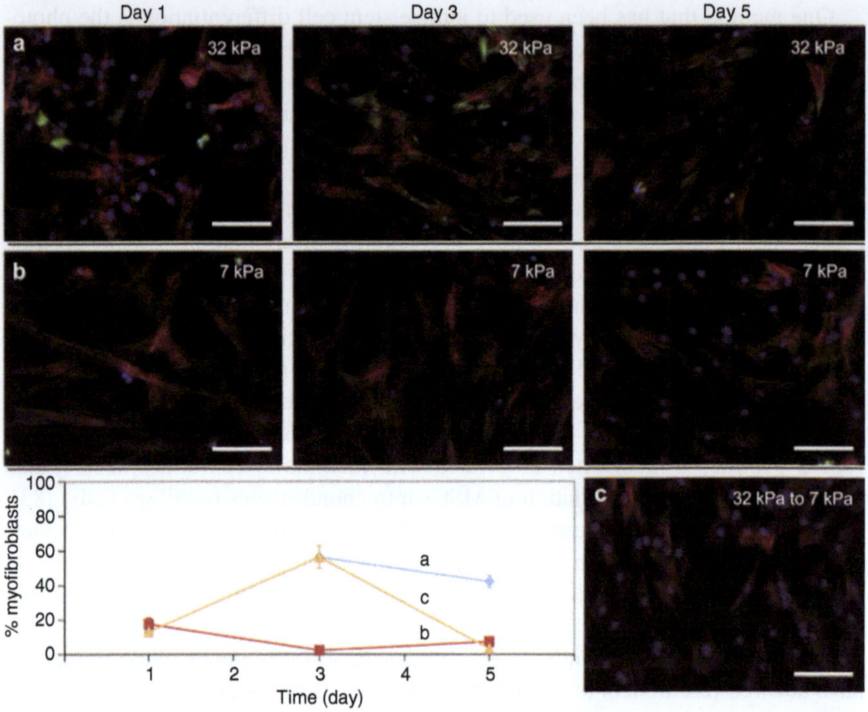

Fig. 4.18 These results demonstrate that actively decreasing substrate stiffness with UV light during culture can cause myofibroblasts to deactivate into VIC cells. VICs were cultured on myofibroblast promoting ((**a**) *E* ~32 kPA) or suppressing ((**b**) *E* ~7 kPa) substrates for 5 days. Cells were also cultured on 32 kPA substrates that were irradiated on day 3 to cause the stiffness to decrease to 7 kPA for the remaining 2 days of culture (**c**). On the dynamic substrate (**c**), nearly all of the cells were deactivated as a result of the decrease in modulus. Scale bars represent 100 μm (*Source*: Reprinted from *Biomaterials*, 31, Kloxin et al., In situ elasticity modulation with dynamic substrates to direct cell phenotype, 1–8, Copyright (2010), with permission from Elsevier)

that contained photo-cleavable cross-links, the hydrogel of this study contained photoisomerizable azobenzene cross-links. Under UV irradiation, the azobenzene in the cross-linker isomerizes into its cis form and the gel is softened. However, under subsequent visible light irradiation, the azobenzene transforms back into its trans isomer and the gel regains stiffness. The changes in stiffness are relatively small (10–100 Pa) compared to the stiffness changes of the nitrobenzene-containing form of the gel. Therefore, when VIC cells were seeded onto the surface and the substrate modulus was switched between the stiffer and softer state via UV irradiation, very little change in phenotype was observed. However, if a larger range of gel stiffness could be achieved by modifying the hydrogel chemistry in the future, this platform has the potential to actively switch VICs back and forth between the activated and deactivated phenotypes.

Another interesting finding regarding stem cell differentiation that has been made with the help of photodegradable hydrogels is that human MSCs posses mechanical memory and can be biased toward osteogenic (bone cell) differentiation when cultured on supraphysiologically stiff materials, such as tissue culture polystyrene (E ~3 GPa), prior to seeding on softer surfaces [86]. To determine if MSCs exhibit mechanical memory, MSCs were cultured on initially stiff hydrogels with a modulus capable of inducing osteogenic differentiation (E ~10 kPA) for either 1, 7, or 10 days (Fig. 4.19a). After this point, the hydrogels were irradiated with UV light

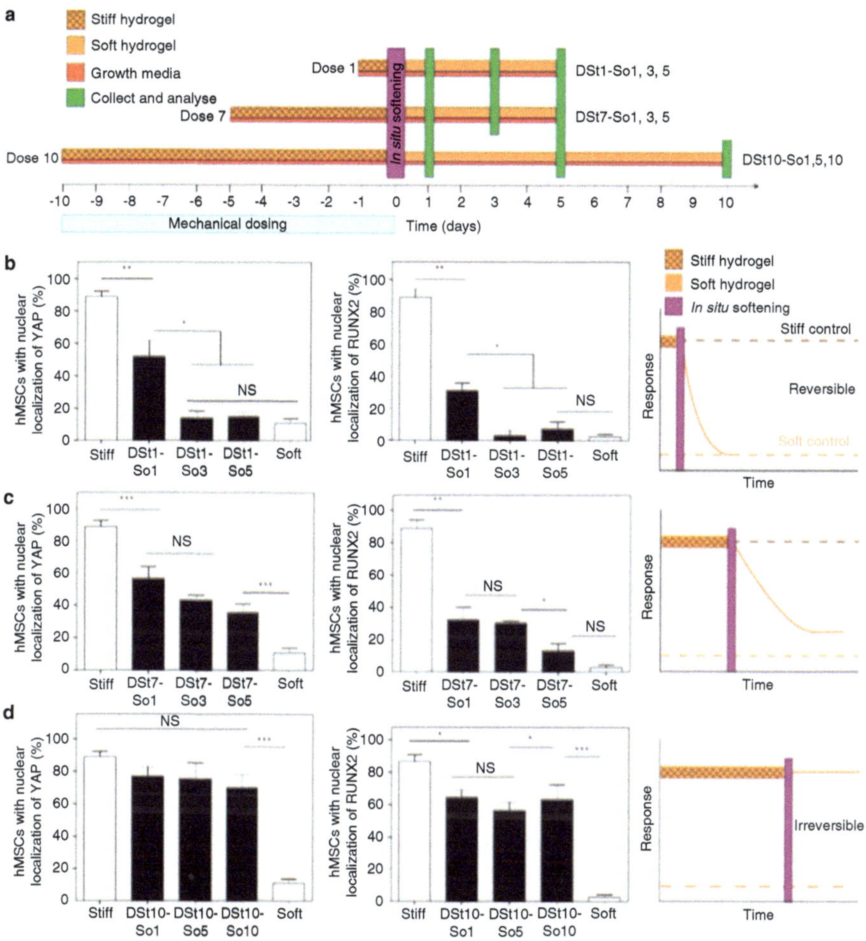

Fig. 4.19 (**a**) MSCs were cultured on stiff hydrogels for 1, 7, or 10 days and then irradiated with UV light to decrease the gel stiffness. YAP and RUNX2 activation were analyzed at 1, 3, 5, and 10 days postirradiation as described in the image. (**b**) The effects of 1 day of mechanical dosing were reversible, as YAP and RUNX2 activation decreased back down to levels of the soft control. (**c**) The effects of 7 days of mechanical dosing were partially reversible, as YAP and RUNX2 activation decreased, but did not reach levels as low as the soft control, even 5 days after softening the gel. (**d**) The effects of 10 days of mechanical dosing were irreversible, as YAP and RUNX2 activation remained significantly higher than the soft control even 10 days after softening the gel (*Source*: Reprinted by permission from Macmillan Publishers Ltd.: [Nature Materials] Yang, C., Tibbitt, M. W., Basta, L. & Anseth, K. S. Mechanical memory and dosing influence stem cell fate. *Nat. Mater.* 13, 645–52 (2014)), Copyright (2014))

to decrease their stiffness to a deactivating modulus (*E* ~2 kPA). The activations of Yes-associated protein (YAP), transcriptional cofactor with PDZ-binding domain (YAZ), and preosteogenic transcription factor (RUNX2) were monitored for 10 days postirradiation to determine if the cells had begun undergoing osteogenic differentiation or if they reverted back into the deactivated state. These markers were used because it is known that YAP and YAZ are present in the cytoplasm of non-differentiated (deactivated state) MSCs; however, they localize to the nucleus during osteogenic differentiation (activated state) [87, 88]. Therefore, in the study of interest, if YAP/YAZ and RUNX2 move to the cytoplasm postirradiation (soft substrate), it indicates that the cells have deactivated to a non-differentiating form. However, if YAP and RUNX2 persist in the nucleus, it indicates that cells have begun osteogenic differentiation and "remember" the stiff gel environment. The authors determined that mechanical dosing is reversible when cultured on a stiff substrate for only 1 day (Fig. 4.19b): YAP and RUNX2 relocated to cytoplasm after 3 days of culture on the softer gel. However, when mechanically dosed on the stiff gel for longer periods of time (10 days), the dosing is irreversible (Fig. 4.19d): the YAP and RUNX2 persisted in the nucleus even after culturing for 10 days on the softer gel. Therefore, it was proven that MSCs do indeed posses mechanical memory when cultured on stiff substrates. It is important to note that these "stiff" substrates were particularly soft compared to typical cell culture substrates such as tissue culture polystyrene (TCPS) (*E* ~10 kPA vs. 3 GPa) yet were still capable of imprinting mechanical memory onto the cells. This finding is important because it brings to light the fact that the amount of time MSCs spend on stiff, TCPS substrates could bias their differentiation toward a specific phenotype. Therefore, this must be taken into consideration when designing cell culture protocol for MSCs.

4.4 The Future of Photo-Activatable Materials

The majority of the studies performed using photo-activatable materials in cell biomechanics and mechanobiology, thus far, have been focused on material development and demonstrating that the newly developed materials have the ability to dynamically influence cell behavior. A relatively small fraction of the studies published in this area have been fully-realized studies of biomechanics or mechanobiology that yield new insight into cell behavior. As a result, there is still a great deal of knowledge that can be gained by using photo-activatable materials in cell biomechanics and mechanobiology applications. For example, the photo-activatable platforms discussed in Sect. 4.3 could be implemented as alternatives to scratch tests for in vitro wound healing assays [89] or to study how cancer cell migration responds to changes in substrate topography without the risk of exposing the cells to damaging heat.

In addition, much of the work completed in this field has been conducted on materials that activate in response to either UV or NIR irradiation. However, materials that activate under visible light irradiation could also prove useful in cell biomechanics and mechanobiology. Visible light is less strongly absorbed by water, a major component of cell culture media, than both NIR and UV light [90]. Therefore, cell culture platforms responsive to visibile light stimuli would allow for triggering substrates through media with reduced risk of hyperthermic damage to cells due to photothermally heating the surrounding media. One way this could be accomplished is by incorporating nanoparticles that absorb visible light into a SMP capable of triggering under cytocompatible conditions. This method should be feasible considering that SMPs have been developed that trigger under NIR irradiation at cytocompatible conditions [46, 59] and that methods exist for preparing silver and gold nanoparticles that absorb in the visible spectrum and that gold nanoparticle-embedded SMPs have been developed that trigger under 532 nm visible light [91]. Therefore, visible light-triggered SMP substrates are a promising prospect for future photo-activatable materials.

In concluding this chapter, it is important to stress the advantages photo-activatable materials hold over other stimuli-responsive materials in biomechanics and mechanobiology applications. First, triggering activatable materials with light allows for precise control over the activation region since light can be focused down to subcellular-sized areas. Therefore, cell attachment sites on SAMs or stiffness of photodegradable hydrogels can be controlled down to the precision of a single cell or subcellular scales at any location on the surface of a 2D substrate or throughout the bulk of a 3D substrate. Second, the intensity of the light can be varied to control the degree of activation. For example, variation in irradiation intensity allows for the creation of stiffness gradients in photodegradable hydrogels [22]. Light irradiation conditions can also be easily controlled in time, allowing for temporal control over the photo-activatable material's properties. Finally, activating stimuli-responsive materials with light irradiation allows for triggering of the material while maintaining high cell viability. The high degree of control over the mechanical properties of photo-activatable substrates makes them excellent candidates for furthering the field of cell biomechanics and mechanobiology.

References

1. D.-H. Kim, P.K. Wong, J. Park, A. Levchenko, Y. Sun, Microengineered platforms for cell mechanobiology. Annu. Rev. Biomed. Eng. **11**, 203–233 (2009)
2. J. Wolff, *Das Gesetz Der Transformation Der Knochen* (A. Hirschwald, Berlin, 1891)
3. J.M. Mitchison, M.M. Swann, The mechanical properties of the cell surface. J. Exp. Biol. **32**, 734–750 (1954)
4. R.M. Hochmuth, Micropipette aspiration of living cells. J. Biomech. **33**, 15–22 (2000)
5. K.L. Sung, M.K. Kwan, F. Maldonado, W.H. Akeson, Adhesion strength of human ligament fibroblasts. J. Biomech. Eng. **116**, 237–242 (1994)
6. M. Radmacher, Measuring the elastic properties of biological samples with the AFM. IEEE Eng. Med. Biol. Mag. **16**, 47–57 (1997)
7. H. Haga et al., Elasticity mapping of living fibroblasts by AFM and immunofluorescence observation of the cytoskeleton. Ultramicroscopy **82**, 253–258 (2000)

8. Y.J. Kim et al., A study of compatibility between cells and biopolymeric surfaces through quantitative measurements of adhesive forces. J. Biomater. Sci. Polym. Ed. **14**, 1311–1321 (2003)
9. O. Thoumine, P. Kocian, A. Kottelat, J. Meister, Short-term binding of fibroblasts to fibronectin: Optical tweezers experiments and probabilistic analysis. Eur. Biophys. J. **29**, 398–408 (2000)
10. R.L.Y. Sah et al., Biosynthetic response of cartilage explants to dynamic compression. J. Orthop. Res. **7**, 619–636 (1989)
11. S. Noria et al., Assembly and reorientation of stress fibers drives morphological changes to endothelial cells exposed to shear stress. Am. J. Pathol. **164**, 1211–1223 (2004)
12. R. Yoshida et al., Comb-type grafted hydrogels with rapid deswelling response to temperature changes. Nature **374**, 240–242 (1995)
13. S. Dai, P. Ravi, K.C. Tam, pH-responsive polymers: synthesis, properties and applications. Soft Matter **4**, 435 (2008)
14. X. Yin, A.S. Hoffman, P.S. Stayton, Poly(N-isopropylacrylamide- co -propylacrylic acid) copolymers that respond sharply to temperature and pH. Biomacromolecules **7**, 1381–1385 (2006)
15. Y. Osada, H. Okuzaki, H. Hori, A polymer gel with electrically driven motility. Nature **355**, 242–244 (1992)
16. S. Tasoglu et al., Guided and magnetic self-assembly of tunable magnetoceptive gels. Nat. Commun. **5**, 4702 (2014)
17. B. Yang, W.M. Huang, C. Li, L. Li, Effects of moisture on the thermomechanical properties of a polyurethane shape memory polymer. Polymer (Guildf). **47**, 1348–1356 (2006)
18. H. Yamaguchi et al., Photoswitchable gel assembly based on molecular recognition. Nat. Commun. **3**, 603 (2012)
19. A. Lendlein, H. Jiang, O. Jünger, R. Langer, Light-induced shape-memory polymers. Nature **434**, 879–882 (2005)
20. A.M. Kloxin, A.M. Kasko, C.N. Salinas, K.S. Anseth, Photodegradable hydrogels for dynamic tuning of physical and chemcial properties. Science **324**, 59–63 (2009)
21. J. Nakanishi et al., Photoactivation of a substrate for cell adhesion under standard fluorescence microscopes. J. Am. Chem. Soc. **126**, 16314–16315 (2004)
22. A.M. Kloxin, M.W. Tibbitt, A.M. Kasko, J.A. Fairbairn, K.S. Anseth, Tunable hydrogels for external manipulation of cellular microenvironments through controlled photodegradation. Adv. Mater. **22**, 61–66 (2010)
23. S.J. Bryant, C.R. Nuttleman, K.S. Anseth, Cytocompatibility of UV and visible light photoinitiating systems on cultured NIH/3T3 fibroblasts in vitro. J. Biomater. Sci. Polym. Ed. **11**, 439–457 (2000)
24. *Biomaterials Science: An Introduction to Materials and Medicine*. (Elsevier Academic Press, New York, 2004)
25. K. Han, W.-N. Yin, J.-X. Fan, F.-Y. Cao, X.-Z. Zhang, Photo-activatable substrates for site-specific differentiation of stem cells. ACS Appl. Mater. Interfaces **7**, 23679–23684 (2015)
26. Y.-H. Gong et al., Photoresponsive 'smart template' via host-guest interaction for reversible cell adhesion. Macromolecules **44**, 7499–7502 (2011)
27. D. Liu, Y. Xie, H. Shao, X. Jiang, Using azobenzene-embedded self-assembled monolayers to photochemically control cell adhesion reversibly. Angew. Chem. Int. Ed. **48**, 4406–4408 (2009)
28. I. Tomatsu, K. Peng, A. Kros, Photoresponsive hydrogels for biomedical applications. Adv. Drug Deliv. Rev. **63**, 1257–1266 (2011)
29. G.D. Nicodemus, S.J. Bryant, Cell encapsulation in biodegradable hydrogels for tissue engineering applications. Tissue Eng. Part B Rev. **14**, 149–165 (2008)
30. M.T. Frey, Y. Wang, A photo-modulatable material for probing cellular responses to substrate rigidity. Soft Matter **5**, 1918–1924 (2009)

31. A.M. Rosales, K.M. Mabry, E.M. Nehls, K.S. Anseth, Photoresponsive elastic properties of azobenzene-containing poly(ethylene-glycol)-based hydrogels. Biomacromolecules **16**, 798–806 (2015)
32. M. Behl, A. Lendlein, Shape-memory polymers. Mater. Today **10**, 20–28 (2007)
33. M. Behl, M.Y. Razzaq, A. Lendlein, Multifunctional shape-memory polymers. Adv. Mater. **22**, 3388–3410 (2010)
34. C. Liu, H. Qin, P.T. Mather, Review of progress in shape-memory polymers. J. Mater. Chem. **17**, 1543 (2007)
35. P.T. Mather, X. Luo, I.A. Rousseau, Shape memory polymer research. Annu. Rev. Mater. Res. **39**, 445–471 (2009)
36. A. Lendlein, R. Langer, Biodegradable, elastic shape-memory polymers for potential biomedical applications. Science **296**, 1673–1676 (2002)
37. W. Small IV, P. Singhal, T.S. Wilson, D.J. Maitland, Biomedical applications of thermally activated shape memory polymers. J. Mater. Chem. **20**, 3356–3366 (2010)
38. K.A. Davis, X. Luo, P.T. Mather, J.H. Henderson, Shape memory polymers for active cell culture. J. Vis. Exp. (2011). https://doi.org/10.3791/2903
39. R.M. Baker, J.H. Henderson, P.T. Mather, Shape memory poly(ε-caprolactone)-co-poly(ethylene glycol) foams with body temperature triggering and two-way actuation. J. Mater. Chem. B **1**, 4916–4920 (2013)
40. L.F. Tseng, P.T. Mather, J.H. Henderson, Shape-memory-actuated change in scaffold fiber alignment directs stem cell morphology. Acta Biomater. **9**, 8790–8801 (2013)
41. H. Lv, J. Leng, Y. Liu, S. Du, Shape-memory polymer in response to solution. Adv. Eng. Mater. **10**, 592–595 (2008)
42. H. Koerner, G. Price, N.A. Pearce, M. Alexander, R.A. Vaia, Remotely actuated polymer nanocomposites—Stress-recovery of carbon-nanotube-filled thermoplastic elastomers. Nat. Mater. **3**, 115–120 (2004)
43. N.G. Sahoo, Y.C. Jung, J.W. Cho, Electroactive shape memory effect of polyurethane composites filled with carbon nanotubes and conducting polymer. Mater. Manuf. Process. **22**, 419–423 (2007)
44. D.J. Maitland, M.F. Metzger, D. Schumann, A. Lee, T.S. Wilson, Photothermal properties of shape memory polymer micro-actuators for treating stroke. Lasers Surg. Med. **30**, 1–11 (2002)
45. W. Small IV, T.S. Wilson, W.J. Benett, J.M. Loge, D.J. Maitland, Laser-activated shape memory polymer intravascular thrombectomy device. Opt. Express **13**, 8204–8213 (2005)
46. Q. Shou, K. Uto, M. Iwanaga, M. Ebara, T. Aoyagi, Near-infrared light-responsive shape-memory poly(ε-caprolactone) films that actuate in physiological temperature range. Polym. J. **46**, 492–498 (2014)
47. Y. Yu, T. Ikeda, Photodeformable polymers: A new kind of promising smart material for micro- and nano-applications. Macromol. Chem. Phys. **206**, 1705–1708 (2005)
48. A. Lendlein, M. Behl, B. Hiebl, C. Wischke, Shape-memory polymers as a technology platform for biomedical applications. Expert Rev. Med. Dev. **7**, 357–379 (2010)
49. K.A. Davis, K.A. Burke, P.T. Mather, J.H. Henderson, Dynamic cell behavior on shape memory polymer substrates. Biomaterials **32**, 2285–2293 (2011)
50. R.M. Baker, L.F. Tseng, M.T. Iannolo, M.E. Oest, J.H. Henderson, Self-deploying shape memory polymer scaffolds for grafting and stabilizing complex bone defects: A mouse femoral segmental defect study. Biomaterials **76**, 388–398 (2016)
51. X. Xu et al., Shape memory RGD-containing networks: Synthesis, characterization, and application in cell culture. Macromol. Symp. **309–310**, 162–172 (2011)
52. P. Yang, R.M. Baker, J.H. Henderson, P.T. Mather, In vitro wrinkle formation via shape memory dynamically aligns adherent cells. Soft Matter **9**, 4705–4714 (2013)
53. R.M. Baker, M.E. Brasch, M.L. Manning, J.H. Henderson, Automated, contour-based tracking and analysis of cell behaviour over long time scales in environments of varying complexity and cell density. J. R. Soc. Interface **11**, 20140386 (2014)

54. M. Ebara et al., Focus on the interlude between topographic transition and cell response on shape-memory surfaces. Polym. (United Kingdom) **55**, 5961–5968 (2014)
55. T. Gong et al., The control of mesenchymal stem cell differentiation using dynamically tunable surface microgrooves. Adv. Healthc. Mater. **3**, 1608–1619 (2014)
56. P.Y. Mengsteab et al., Spatiotemporal control of cardiac anisotropy using dynamic nanotopographic cues. Biomaterials **86**, 1–10 (2016)
57. S.A. Turner, J. Zhou, S.S. Sheiko, V.S. Ashby, Switchable micropatterned surface topographies mediated by reversible shape memory. ACS Appl. Mater. Interfaces **6**, 8017–8021 (2014)
58. D.M. Le, M.A. Tycon, C.J. Fecko, V.S. Ashby, Near-infrared activation of semi-crystalline shape memory polymer nanocomposites. J. Appl. Polym. Sci. **130**, 4551–4557 (2013)
59. Q. Shou, K. Uto, W.-C. Lin, T. Aoyagi, M. Ebara, Near-infrared-irradiation-induced remote activation of surface shape-memory to direct cell orientations. Macromol. Chem. Phys. **215**, 2473–2481 (2014)
60. C.A. Goubko, S. Majumdar, A. Basak, X. Cao, Hydrogel cell patterning incorporating photocaged RGDS peptides. Biomed. Microdev. **12**, 555–568 (2010)
61. J. Nakanishi et al., Spatiotemporal control of cell adhesion on a self-assembled monolayer having a photocleavable protecting group. Anal. Chim. Acta **578**, 100–104 (2006)
62. J. Nakanishi et al., Spatiotemporal control of migration of single cells on a photoactivatable cell microarray. J. Am. Chem. Soc. **129**, 6694–6695 (2007)
63. J. Nakanishi, H. Nakayama, K. Yamaguchi, A.J. Garcia, Y. Horiike, Dynamic culture substrate that captures a specific extracellular matrix protein in response to light. Sci. Technol. Adv. Mater. **12**, 44608 (2011)
64. Y. Kikuchi et al., Grafting poly(ethylene glycol) to a glass surface via a photocleavable linker for light-induced cell micropatterning and cell proliferation control. Chem. Lett. **37**, 1062–1063 (2008)
65. Y. Kikuchi et al., Arraying heterotypic single cells on photoactivatable cell-culturing substrates. Langmuir **24**, 13084–13095 (2008)
66. S. Kaneko et al., Photocontrol of cell adhesion on amino-bearing surfaces by reversible conjugation of poly(ethylene glycol) via a photocleavable linker. Phys. Chem. Chem. Phys. **13**, 4051–4059 (2011)
67. M. Kamimura et al., Facile preparation of a photoactivatable surface on a 96-well plate: A versatile and multiplex cell migration assay platform. Phys. Chem. Chem. Phys. **17**, 14159–14167 (2015)
68. S. Petersen et al., Phototriggering of cell adhesion by caged cyclic RGD peptides. Angew. Chem. Int. Ed. **47**, 3192–3195 (2008)
69. J. Nakanishi et al., Precise patterning of photoactivatable glass coverslip for fluorescence observation of shape-controlled cells. Supramol. Chem. **22**, 396–405 (2010)
70. Y. Shimizu, H. Boehm, K. Yamaguchi, J.P. Spatz, J. Nakanishi, A photoactivatable nanopatterned substrate for analyzing collective cell migration with precisely tuned cell-extracellular matrix ligand interactions. PLoS One **9**(3), e91875 (2014). https://doi.org/10.1371/journal.pone.0091875
71. Y. Ohmuro-Matsuyama, Y. Tatsu, Photocontrolled cell adhesion on a surface functionalized with a caged arginine-glycine-aspartate peptide. Angew. Chem. Int. Ed. **47**, 7527–7529 (2008)
72. M.D. Pierschbacher, E. Ruoslahti, Cell attachment activity of fibronectin can be duplicated by small synthetic fragments of the molecule. Nature **309**, 30–33 (1984)
73. G.S. Nowakowski et al., A specific heptapeptide from a phage display peptide library homes to bone marrow and binds to primitive hematopoietic stem cells. Stem Cells **22**, 1030–1038 (2004)
74. F.-Y. Cao, W.-N. Yin, J.-X. Fan, R.-X. Zhuo, X.-Z. Zhang, A novel function of BMHP1 and cBMHP1 peptides to induce the osteogenic differentiation of mesenchymal stem cells. Biomater. Sci. **3**, 345–351 (2015)

75. M.W. Tibbitt, A.M. Kloxin, K.U. Dyamenahalli, K.S. Anseth, Controlled two-photon photodegradation of PEG hydrogels to study and manipulate subcellular interactions on soft materials. Soft Matter **6**, 5100 (2010)
76. B. Wildt, D. Wirtz, P.C. Searson, Programmed subcellular release for studying the dynamics of cell detachment. Nat. Methods **6**, 211–213 (2009)
77. A.J. Ridley et al., Cell migration: Integrating signals from front to back. Science **302**, 1704–1709 (2003)
78. P. Friedl, B. Weigelin, Interstitial leukocyte migration and immune function. Nat. Immunol. **9**, 960–969 (2008)
79. P.L. Ryan, R.A. Foty, J. Kohn, M.S. Steinberg, Tissue spreading on implantable substrates is a competitive outcome of cell-cell vs. cell-substratum adhesivity. Proc. Natl. Acad. Sci. **98**, 4323–4327 (2001)
80. J. Bourget, M. Guillemette, T. Veres, F.A. Auger, L. Germain, Alignment of cells and extracellular matrix within tissue-engineered substitutes. Adv. Biomater. Sci. Biomed. Appl. Ref., 365–390 (2013). https://doi.org/10.5772/54142
81. C.M. Kirschner, D.L. Alge, S.T. Gould, K.S. Anseth, Clickable, photodegradable hydrogels to dynamically modulate valvular interstitial cell phenotype. Adv. Healthc. Mater. **3**, 649–657 (2014)
82. F. Guilak et al., Control of stem cell fate by physical interactions with the extracellular matrix. Cell Stem Cell **5**, 17–26 (2009)
83. S. Tavella et al., Regulated expression of fibronectin, laminin and related integrin receptors during the early chondrocyte differentiation. J. Cell Sci. **110**, 2261–2270 (1997)
84. A.M. Kloxin, J.A. Benton, K.S. Anseth, In situ elasticity modulation with dynamic substrates to direct cell phenotype. Biomaterials **31**, 1–8 (2010)
85. H. Wang, S.M. Haeger, A.M. Kloxin, L.A. Leinwand, K.S. Anseth, Redirecting valvular myofibroblasts into dormant fibroblasts through light-mediated reduction in substrate modulus. PLoS One **7** (2012)
86. C. Yang, M.W. Tibbitt, L. Basta, K.S. Anseth, Mechanical memory and dosing influence stem cell fate. Nat. Mater. **13**, 645–652 (2014)
87. S. Dupont et al., Role of YAP/TAZ in mechanotransduction. Nature **474**, 179–183 (2011)
88. G. Halder, S. Dupont, S. Piccolo, Transduction of mechanical and cytoskeletal cues by YAP and TAZ. Nat. Publ. Gr. **13**, 591–600 (2012)
89. M.J. Salierno, A.J. Garcia, A. Del Campo, Photo-activatable surfaces for cell migration assays. Adv. Funct. Mater. **23**, 5974–5980 (2013)
90. R.M. Pope, E.S. Fry, Absorption spectrum (340–640 nm) of pure water. I. Photothermal measurement. Appl. Opt. **36**, 8710–8723 (1997)
91. H. Zhang, H. Xia, Y. Zhao, Optically triggered and spatially controllable shape-memory polymer–gold nanoparticle composite materials. J. Mater. Chem. **22**, 845–849 (2012)

Chapter 5
Photonics in Drug Delivery

Anna Karewicz, Dorota Lachowicz, and Aneta Pietraszek

5.1 Introduction

Drug delivery is a relatively young branch of pharmacy, focused on designing of the advanced carrier-drug systems in search for more efficient and safe therapy. Lately such systems started to be more widely applied in order to deliver photosensitive agents. One of the reasons may be the fact that light-induced therapies are difficult to be applied inside the human body. Only recently, due to the technical advances, fluorescence-emitting probes and photosensitizers could be applied with success to in vivo diagnosis and treatment of certain diseases, e.g. cancer. Here we would like to give an overview of the drug delivery systems based on photoresponsive properties of some compounds, which were tailored to deliver active agents to their place of action in the human body. These systems could be used either for diagnosis, in order to study the pharmacokinetics of the new drug (its distribution in the body and possible routes of elimination) or to fight the disease.

Among drug delivery systems (DDS) carrying photosensitive compounds, we would like first to point out to the possibility of using fluorescent labels in order to study absorption, distribution, metabolism and elimination of new ones. Part of the chapter will also be dedicated to various DDS designed to deliver photoactive compounds, in particular photosensitizers for photodynamic therapy (PDT) and photogeneration of nitric oxide in the treatment of cancer. Finally different mechanisms for achieving photo-controlled release of the bioactive compound from DDS are being discussed in order to show both advantages and limitations of each approach.

A. Karewicz (✉) · A. Pietraszek
Department of Chemistry, Jagiellonian University, Cracow, Poland
e-mail: karewicz@chemia.uj.edu.pl; aneta.pietraszek@doctoral.uj.edu.pl

D. Lachowicz
AGH University of Science and Technology, Cracow, Poland
e-mail: dbielska@agh.edu.pl

J. Van Hoorick et al. (eds.), *Polymer and Photonic Materials Towards Biomedical Breakthroughs*, Micro- and Opto-Electronic Materials, Structures, and Systems, https://doi.org/10.1007/978-3-319-75801-5_5

5.2 Delivery of Photo-Imaging Agents for Pharmacokinetic Studies

The studies of the pharmacokinetics of the biologically active compound are crucial for verifying the possibility of its application as a drug. They also guide future animal and clinical studies in order to select suitable dose levels and frequency of administration. Among various methods used to perform these studies, a significant part constitutes photo-imaging techniques.

They allow to follow the absorption and distribution of the bioactive agent after introduction to the body in non-invasive way, thus avoiding multiple blood sampling (frequent blood loss can cause trauma and adverse changes in physiology interfering with the obtained results) or sacrifice of animals done in order to control changes in different organs in various time points of the therapy. One of such methods is near-infrared (NIR) optical imaging. Using this method the mapping of vasculature and tumour can be done, as well as metabolic analysis. Dobosz et al. [1] described non-invasive studies of pharmacokinetics using the combination of the fluorescence imaging of the whole mouse's body with the NIR imaging done exclusively on highly vascularized retina of the eye. Using this methodology it was possible to study the pharmacokinetics of OsteoSense 750®, a fluorescent bisphosphonate targeting bone tissue and therapeutic antibody trastuzumab labelled with Alexa 750. It was shown that, in contrast to the conventional techniques, this new methodology allowed to perform pharmacokinetics studies more efficiently, as quantification of the drug in serum and data interpolation between time points was not necessary. Additionally, the number of required mice was significantly reduced, and there was no need to sacrifice the animals.

Fluorescence-based studies of the pharmacokinetics of eribulin mesylate, a potent cytotoxic agent, gave an insight into the possible reason of its failure as an alternative in paclitaxel-resistant cancers [2].

An analog of eribulin with pharmacokinetic and cytotoxic activity across a human cell line panel similar to the parent drug was labelled with boron-dipyrromethene (BODIPY) fluorophore. In vivo confocal microscopy of the anesthetized mice allowed to determine the critical determinants of limited drug accumulation in tumour cells, namely, multidrug resistance protein 1 (MDR-1)-mediated drug efflux and three-dimensional tumour vascular architecture. As a confirmation it was shown that MDR-1 inhibitor (HM30181) encapsulated in nanoparticulate system allowed for efficient accumulation of eribulin mesylate, whereas free inhibitor could not achieve that goal.

The most advanced strategies using optical tools to study and control cells' activity in response to pathology and drug administration are employed in the research on brain functioning, mainly by identifying therapeutic targets in the psychiatric disorders [3]. Optogenetics is a technique of monitoring neurons' activity by introducing light-sensitive ion channels (e.g. channelrhodopsin-2 or halorhodopsin) into neurons. It also allows to regulate the activity of these channels and other light-sensitive proteins in the brain. As an example, the optogenetic investigation of the ionic mechanisms, focused on the phasic firing phenomenon, may reveal novel drug targets for the treatment of major depressive disorder. The way the neuron transmits the signal is by the short-lasting change in the electrical membrane potential of a cell, known as action potential. The emission of the action potential by a neuron is described as "firing". Ventral tegmental area dopamine neurons in the brain reward circuit exhibit two different firing patterns: low-frequency tonic and high-frequency phasic [4]. The phasic firing encodes reward dopamine signals. Based on clear evidence from the phasic optogenetic activation experiments the hyperpolarization-activated cation channels (I_h) were selected as the possible targets of depression therapy. Experimental studies showed that the local infusion of I_h inhibitors (ZD7288 and DK-AH269) gave antidepressant effects within a few hours [5]. Even more importantly, the antidepressant effects induced by a single-dose lasted at least 2 weeks, which is very different from traditional antidepressant medications that take weeks to reach clinical efficacy.

5.3 Delivery of Photoactive Drugs

5.3.1 Photodynamic Therapy (PDT)

PDT is a non-invasive therapeutic approach to treat cancer. It is based on the application of the photosensitive compound, which, upon irradiation with the light of appropriate wavelength, transmits the energy to other molecules, producing cytotoxic spices (e.g. singlet oxygen, $^1O_2$), which induce cell death. Only the cells in the irradiated tissue will be exposed to the cytotoxic effect of PDT (see Fig. 5.1). However, because most of the photosensitizers absorb in the visible region of the spectrum, the rest of the body usually has to be protected from any light until the photosensitive compound is degraded and removed from the organism. Other disadvantages concern the nature of most photosensitizers. These are rather large aromatic molecules (porphyrins, phthalocyanines) which are sparingly soluble in aqueous media and have a strong tendency to aggregate, which leads to the losing their PDT efficiency. To solve these problems, various DDS have been proposed.

5.3.1.1 Micelles

The micelles which are used as DDS most often consist of amphiphilic block or graft copolymers, which are either directly dissolved in water or introduced with miscible organic co-solvent [6]. They are especially suitable for the delivery of hydrophobic compounds. The photosensitizer can be physically entrapped in the hydrophobic core, or, in order to prevent both aggregation and dark cytotoxicity (toxicity with no irradiation involved), it can be covalently bound to the polymeric chain, forming conjugate [7]. The physical encapsulation can be realized via electrostatic interactions (polyion complexes), by evaporation of the organic solvent or by application of the emulsification procedures. A number of polymers have been used

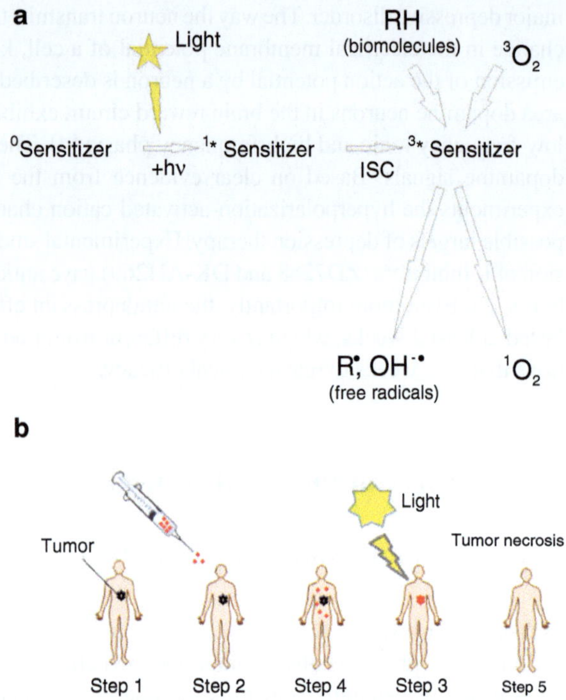

Fig. 5.1 (**a**) The basic sequence of photophysical and photochemical processes leading to the formation of cytotoxic species (free radicals, singlet oxygen) in PDT, (**b**) the scheme of typical PDT procedure

as a hydrophobic part (core), including polyesters [8] poly(amino acids) [9], and poly(acrylamides) [10]. The hydrophilic part, which forms outer shell/corona of the micelle, is built almost exclusively from poly(ethylene glycol) (PEG). It is a neutral and highly biocompatible polymer, which is also characterized by excellent solubility in water. Its main role is to maintain the stability of the micelle while protecting the incorporated drug from enzymes and other proteins present in blood. PEGs with a molecular weight of 1000–12,000 g mol^{-1}and chain length which is greater than or equal to that of hydrophobic chain are usually employed.

Micelles are dynamic systems, and they tend to get disrupted when diluted in the bloodstream below their critical micelle concentration. There are various approaches to solve this problem, including chemical cross-linking of the polymeric chains and physical interactions (e.g. electrostatic interactions or metal-acetate complexation [11]. The introduction of environmentally sensitive polymeric units or cross-linkers enables micelles to respond either to stimuli associated with the local microenvironment of the tumour (e.g. pH, heat) or to the exogenously applied stimuli (e.g. magnetic field) by releasing their photoactive payload [12].

The majority of photosensitizers are characterized by some toxicity towards healthy cells; therefore it is the role of the micelle to safely deliver these agents to the tumour tissue via either passive or active transport. Due to their relatively low, compared to liposomes and nanoparticles, size (in the range of tens of nm) and high uniformity of the size, micelles are best suited to accomplish passive targeting to tumours by the so-called enhanced permeation and retention (EPR) effect [13]. Thanks to the presence of shielding PEG corona they can efficiently avoid immune system achieving prolonged circulation in blood. There is also a possibility to modify the peripheral chain ends of the polymers with ligands or antibodies actively recognized by the cells of diseased tissue, achieving active targeting of the photosensitizer. One example is that of phthalocyanine encapsulated within poly(caprolactone)-block-PEG copolymer micelles [14]. The building copolymer units were conjugated with small peptide having high affinity to epidermal growth factor receptor, which is significantly upregulated in several cancers.

There is a variety of nanoparticulate systems designed to deliver sensitizers for PDT [15]. Some of the earlier described micellar systems, once chemically cross-linked, may as well be described as nanoparticles. Most of them are based on polymers, including polyesters such as poly(lactide) (PLA), poly(lactide-*co*-glycolide (PLGA) or poly(ethylene glycol-*co* lactide (PEG-PLA) [16]. There are, however, groups of nanoparticulate systems with a very distinctive characteristics and often inorganic or hybrid inorganic-organic nature, which have been used with success in PDT. They will be discussed in the following sections of this chapter.

5.3.1.2 Layered Double Hydroxide (LDH) Nanoparticles

LDH are mineral and synthetic materials with a structure consisting of positively charged layers of mixed metal hydroxides. Exchangeable anions, compensating the charge, are located in the interlayer spaces [17]. The ion-exchange properties allow for the introduction of the ionically modified photoactive compounds and render control over their subsequent release. LDH have positive surface charge, which facilitates internalization with cell membranes. As a drug delivery system, LDH material has been shown to increase biocompatibility, bioavailability [18] and solubility of hydrophobic drugs. The use of LDH for PDT has been studied only very recently, but these structures show great potential [19]. For example, the LDH particles with intercalated near-infrared photoactive dye, indocyanine green, were coated with chitosan and tested for the drug stability and photodynamic effect [20]. The nanoparticles were fully internalized into the lysosomes of HT-29 human colorectal adenocarcinoma cells after 4 h of incubation. The intercalation of the photosensitizer caused some aggregation, but due to the shielding effect of LDH, the system allowed for the enhanced PDT effect in vitro. LDH were tested with success for several PDT systems including polyurethane composites [21] and pH-responsive nanohybrid materials [22].

5.3.1.3 Silica Nanoparticles

Although silica-based particles are not biodegradable, neither possess any special nanosize-related properties, they also exhibit a number of advantages for the delivery of PDT agents. Silica materials are biocompatible and may be organized into the highly cross-linked, relatively rigid structure, which allows for obtaining mesoporous particles [23], nanocapsules and multishell systems [24]. The preparation process can be performed in ambient temperature, and the obtained particles are not susceptible to changes in pH (e.g. do not swell) nor vulnerable to microbial or any enzymatic attack as, for example, hydrogel particles. While typically in drug delivery, this lack of responsiveness may be considered a disadvantage, as the particle should be able to release the loaded drug; it is not mandatory when photosensitizer is being delivered, as long as the matrix is penetrable by light and molecular and singlet oxygen. Roy and co-workers [25] have developed the photosensitizer-doped, organically modified silica nanoparticles with the average size of 30 nm by controlled hydrolysis of triethoxyvinylsilane (TEVS) in micellar media. 2-Devinyl-2-(1-hexyloxyethyl) pyropheophorbide (HPPH) was used as a photosensitizer. HPPT-containing nanoparticles were able to efficiently generate singlet oxygen. In vitro studies have demonstrated the active uptake of this nanoparticulate system into the cytosol of tumour cells, and significant death of such cells was observed upon irradiation ($\lambda = 650$ nm).

Even more interesting are materials based on organically modified silica, obtained using organosilane precursors. Such modifications allow for the introduction of a variety of organic moieties, both on the surface of the nanoparticles and into their internal structure. As a result various functionalities can be incorporated, and the properties of the particles may be fine-tuned. A good example of such a system was proposed by Selvestrel and co-workers [26]. The nanoparticles were obtained by the ammonia-catalysed polymerization of alkoxysilane precursor in water in the presence of surfactant. The photosensitizer was introduced in the form of alkoxysilane derivative of porphyrin (temoporfin, mTHPC). In order to prevent interaction with serum proteins, the nanoparticles' surface was densely pegylated by addition of an amphiphilic trialkoxysilane PEG derivative. Surface amine groups were introduced by the addition of amine-PEG-silane derivative in order to decorate the nanoparticles with targeting moieties (biotin, folic acid or cetuximab antibodies). It was found that the organosilica matrix provides a suitable environment to preserve the photochemical properties of the photosensitizer. Pegylation decreases dark cytotoxicity of the system and minimizes RES capture but also hinders cell uptake. It can be though easily overcome by using well-chosen targeting structure (in this case cetuximab antibodies).

5.3.1.4 Virus Capsids: Targeted Delivery

The idea of using virus capsids or, more generally, virus-like particles in biomedical applications has been intensely exploited over the last decade [27]. They are highly biocompatible and have developed with the aim to enter the cell. Therefore they are characterized by the optimal size and high stability and can utilize cell mechanisms to cross cell membrane. Additionally a number of strategies have been developed in order to modify the surface of virus capsids/particles according to the application [28]. Cohen and Bergkvist [29] have reported the targeted PDT system based on MS2 bacteriophage capsid. As a photosensitizing molecule, the porphyrin, meso-tetra (4-N,N,N-trimethylanilinium)-porphine (TMAP), was used. Due to its positive charge, TMAP could be easily loaded inside the capsid. Its exterior was decorated with a DNA aptamer (short sequence of DNA) with repeating GGT motifs, which are known to bind to nucleolin, the protein overexpressed on the surface of many types of cancer cells [30]. The authors have shown that the system binds specifically to the MCF-7 breast cancer cells and induces cytotoxicity upon irradiation. There was no toxic effect observed on the control MCF-10 cell line. It was also confirmed that phototoxicity of the system was directly related to the targeting ability of the aptamer-modified capsid.

5.3.1.5 Protease-Sensitive Prodrugs

Proteases are a group of enzymes which play an important role in a number of key biological processes including cell proliferation, migration and death. Although normally they are strictly controlled, their modulated activity is characteristic for many neoplastic and invasive diseases. One of the ways to take advantage of the up-regulation of certain proteases in PDT involve the development of drug delivery systems designed to be selectively converted from non-toxic prodrug to an active photosensitizer by proteolysis. If the carrier is built of the polymeric chain, the system may show selectivity towards cancer cells on two distinct levels: alterations in the proteolytic activity and changes in the structure of the diseased vasculature [31]. Optimally, the photoactive compound is optically inert in the form of a prodrug, while its photosensitizing properties are restored when released by the target protease. This allows to minimize side-effects. In many cases the photosensitizing unit can be also used as fluorescence imaging agent, yielding dual-modality prodrug which can both monitor and treat protease-overexpressing cells upon irradiation.

The whole class of fluorescence imaging systems is based on protease-sensitive polymeric backbones with photoactive moieties attached to it. The relatively high number of photoactive units on one polymeric chain allows for aggregation-driven quenching of their fluorescence, thus providing low background signal. Upon fragmentation of the backbone induced by proteases, the chromophores get separated, and their fluorescence is restored. One of such systems consists of poly(L-lysine) backbone with near-infrared fluorescence probe (Cy5.5) densely grafted via short methoxy-PEG chains [32]. This conjugate resulted in 12-fold increase in fluorescence intensity after protease-mediated activation in vitro and a bright signal from the tumour in vivo with the ratio of diseased to healthy tissue of approximately seven. The same conjugate was successfully applied in a model of rheumatoid arthritis.

Although such systems show good efficacy, they rely on the proteolytic cleavage of the polymeric backbone and therefore may not provide sufficient specificity for a particular protease. Another step in the development of such systems led to the conjugates in which the fluorescent agent is bound to the proteolysis-resistant polymeric chain via the linker sensitive to a specific protease [33].

The imaging systems similar to those described above can be applied to PTD when the fluorescent agent will be replaced by a photosensitizer. Such protease-sensitive prodrugs are highly specific to diseased tissue due to the presence of disease-specific trigger. Due to their sufficiently high fluorescence quantum yields, such prodrug can still be used for imaging. Covalent attachment of polyethylene glycol to macromolecules may alter their pharmacokinetics, targeting ability and photophysical properties.

Hamblin and co-workers [34] have shown the differences in the influence of pegylation on the properties of the cationic and anionic conjugates of poly-L-lysine with a photosensitizer chlorin$_{e6}$. The cationic conjugate after pegylation became less aggregated, consumed less oxygen and had reduced cellular uptake. Its phototoxicity (corrected for cellular uptake) increased three- to fivefold. In contrast, the anionic conjugate on pegylation became more aggregated and consumed similar amounts of oxygen and cellular uptake, while its relative phototoxicity was reduced. It was shown that a single low dose of PDT (0.125 mg chlorin$_{e6}$ equivalent/kg) poly-L-lysine-chlorin$_{e6}$ conjugate grafted with methoxypolyethylene glycol was sufficient to suppress HT1080 human fibrosarcoma tumour growth by more than 50% [35]. Importantly the same concept can be used to obtain PDT prodrugs sensitive to any functional enzyme. One example is the pullulan conjugate containing multiple photosensitizing pheophorbide units and targeted to foliate receptors. After internalization into the cancer cells, the photosensitizer was activated by esterases in the lysosomes [36].

5.3.1.6 Curcumin-Metal Complexes in PDT

Curcumin, 1,7-bis(4-hydroxy-3-methoxyphenyl)-1,6-heptane-3,5-dione (diferuloylmethane), is a lipophilic natural polyphenol derived from the rhizome of turmeric (*Curcuma longa*). It is a photoactive compound and its photobiological activity is based on the ability to generate reactive oxygen species (ROS). Curcumin molecule in the triplet state absorbs at 720 nm and reacts with molecular oxygen to generate $^1O_2$. The photoinduced ROS generation by curcumin leads to cancer cell death. Curcumin potential to be used as cytostatic and anticancer drug is, however, severely limited by its poor bioavailability and lack of hydrolytic stability in biological media.

Recently it was shown that when complexed with metal ions, curcumin does not lose its photoactivity, while its hydrolytic stability and solubility in the aqueous media increased considerably [37]. Various complexes have been studied so far in search to increase bioavailability, stability and anticancer activity of curcumin, including zinc [38], cobalt [39], ruthenium [40], copper [41], lanthanum [42] and oxovanadium [43]. Among them complexes with oxovanadium(IV) and copper(II) gave best results with increased hydrolytic stability and the photocytotoxic effect enhanced ca. five times compared to curcumin alone. The ternary oxovanadium(IV) complex of curcumin, having (acridinyl)dipyridophenazine (acdppz) ligand, was found to be remarkably photocytotoxic upon visible light irradiation (IC50 of 0.7 μM in HeLa cells) while being significantly less toxic in darkness (IC50 > 10 μM). It also showed mitochondrial localization, preferred in PDT applications [44].

5.3.2 *Light-Triggered Nitric Oxide (NO) Delivery*

Endogenous NO mediates multiple processes in the cardiovascular system, the immune system and the central and peripheral nervous system. A number of potential therapeutic applications of exogenous NO in the treatment of cardiovascular and respiratory disease, cancer treatment, antimicrobial therapy and wound healing have been proposed so far [45]. The main limitations of these therapies are short half-life of nitric oxide and its vulnerability to various biological substances. The versatile functions of NO in human body are closely linked to its ability to convert between various oxidation forms: highly toxic nitrogen oxide, very reactive species, such as dinitrogen trioxide (N_2O_3) or peroxynitrite ($ONOO^-$) or more stable than NO nitrite and nitrate anions. The production of these reactive nitrogen species (RNS) has to be taken into consideration in NO-based therapies, as RNS are able to cause irreversible modifications of the biological molecules, including genetic material. Thus precise control over NO delivery seems critical.

A number of NO-releasing molecules (so-called NO donors) have been synthesized so far, including glyceryl trinitrate, nitrosothiols and N-diazeniumdiolates [46]; however they do not provide sufficient control over the NO release in order to be used for selective destruction of infected or malignant sites via higher concentrations of nitric oxide. As a solution to this problem, the photoactive NO complexes of transition metals (metal nitrosyls) were synthesized by Heilman and Mascharak, designed to deliver NO at selected sites through the control of light exposure [47]. Two series of photoactive ligands derived from pentadentate (PaPy3H) and tetradentate (H_2bpb) structures were developed (H designate dissociable carboxyamide proton) (see Fig. 5.2). Both types of structures were proved effective in the light-triggered delivery

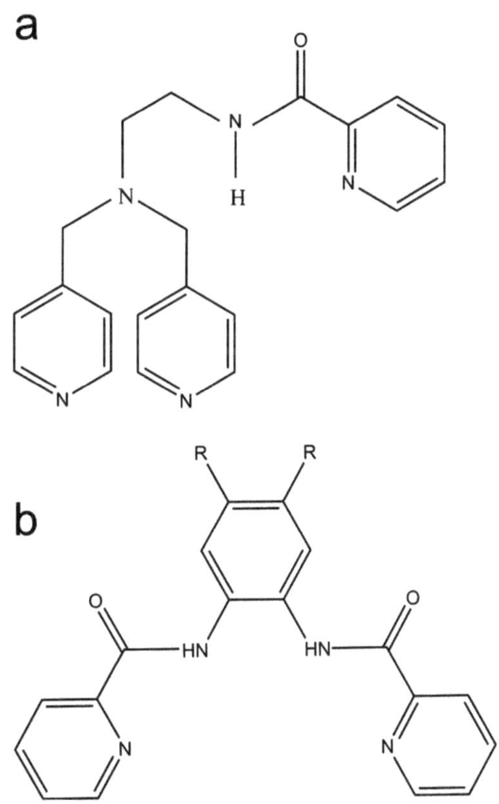

Fig. 5.2 Designed ligand structures used to obtain photoactive metal nitrosyls (**a**) $PaPy_3H$, (**b**) H_2bpb

of NO to biological targets. To achieve that, however, the structures had to be modified with appropriate substituents (to redshift the energy of photo release), as well as by the direct coordination of dyes (to obtain additional photosensitization). They were also combined with polymeric matrix to enable targeting. Light-sensitive polyurethane-based film containing dispersed xerogel particles was the most effective example of targeted light-triggered delivery of NO using these type of delivery systems [48].

5.3.2.1 Mitochondria-Targeted Photoactive NO-Releasing Carbon Dots

Carbon dots are a new class of fluorescent carbon nanomaterials (~ 10 nm in diameter), which show excellent cell membrane permeability, tuneable surface functionalities and good photostability in addition to interesting optical properties. Recently they were used to build the multifunctional system for antitumour therapy [49]. Carbon dots were surface-modified with triphenylphosphonium moiety in order to target mitochondria (organelle responsible for apoptosis) and conjugated with a photoresponsive NO donor (derivative of 4-nitro-3-(trifluoromethyl)aniline). Two molecular NO-releasing systems (one containing targeting ligand and one without the ligand) and two NO-releasing systems based on carbon dots (with and without targeting ligand) were synthesized. This allowed to monitor both release of nitric acid and mitochondrial targeting. Both processes were monitored by following the co-localization of the fluorescent carbon dots-based systems with mitochondria-specific molecular probes: rhodamine 123 (Rh123) and JC-1. The system was shown to exhibit low dark cytotoxicity and considerable cytotoxicity upon irradiation with UV light due to triggered release of NO. It could also effectively target mitochondria, causing photoinduced apoptosis. The main advantage lies in its high compatibility with aqueous media (no need for using organic solvent), while the main drawback reminds the difficulty to accurately characterize its exact structure.

5.4 Photo-Controlled Drug Delivery Systems

Whereas in the previously described systems, light was involved directly in the biological activity of the proposed systems; it may also be used as an external stimuli introduced to control time and place of release of the biologically active species. There is quite a variety of drug delivery systems controlled by light described in literature, including these based on liposomes, polymer hydrogels, prodrugs and hybrid nanoparticles.

5.4.1 Photo-Triggered Release in Liposomal Systems

The photo-triggered release of the biologically active agent may be realized in liposomal systems by using different release mechanisms, including photo-cleavage [50–52], photoisomerization [53], photopolymerization of lipids [54, 55] and photothermal processes [56–58]. The main idea behind each mechanism is shown in Fig. 5.3.

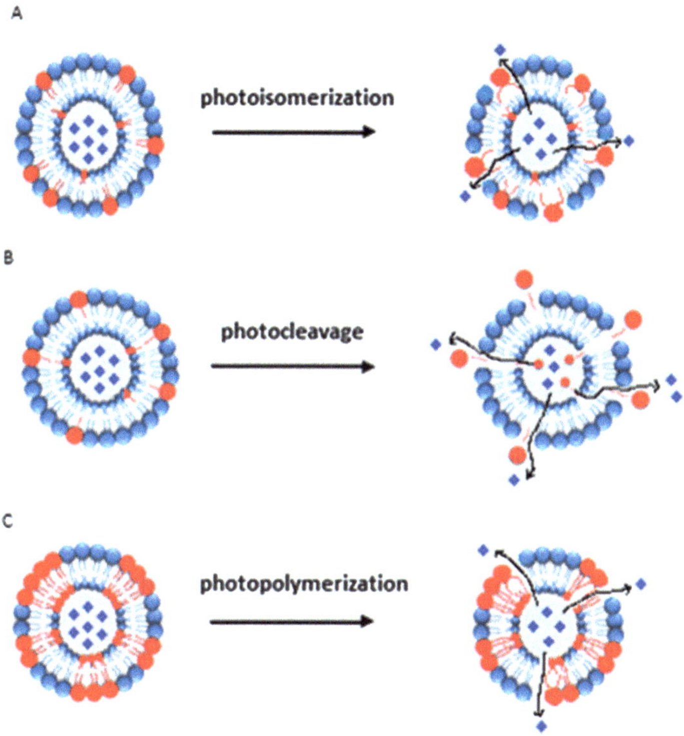

Fig. 5.3 The scheme illustration of the mechanism of release of the bioactive agent from liposomes based on (**a**) photoisomerization, (**b**) photo-cleavage and (**c**) photopolymerization

5.4.1.1 Photo-Cleavable Liposomes

The concept of photo-cleavable liposomes is based on the inclusion of photolabile lipid molecules into the liposomal membrane. The illumination of such vesicles leads to rupture of the labile bonds resulting in separation of the polar and non-polar fragments of the introduced photosensitive molecules. As a consequence, the amphiphilic character of these molecules is lost, and destabilization of the liposomes occurs leading to the release of their content. This strategy of photo-triggered release from liposomes containing photo-cleavable lipids was already reported in the 1990s by Thompson's group [11]. In their work they described the photoinduced cleavage of naturally occurring lipids called plasmalogens built in the bilayer of liposomal vesicles. Liposomes were made in 80% or more from.

plasmalogen in combination with auxiliary lipids, such as lecithin, PEG-lipid or cholesterol. Nowadays, many types of photo-cleavable lipids are known, which were designed specifically to cleave upon irradiation giving the breakdown products which will be the most effective in rapid destabilization of liposome membrane. Some of the examples of such photo-cleavable lipids activated with UV light include NVOC-DOPE (1,2-dioleoyl-sn-phosphatidylcholine modified with 6-nitroveratryloxycarbonyl chloride) [52], amphiphilic lipids built from the stearyl amine non-polar tail and charged amino acid polar head bound together via the o-nitrobenzyl group [50], or phosphocholine derivatives obtained by connecting the hydrophilic phosphocholine with hydrophobic tails via a photolabile, dithiane-based tether [59].

5.4.1.2 Liposomes Triggered by Photoisomerization

The release of a drug from liposomes based on photoisomerization mechanism is achieved by introducing into the liposomal bilayer a molecule which, upon irradiation, undergoes a conformational change. This change destabilizes liposomal membrane and allows for release of the content. Typical photoisomerizable molecules used for controlled release of liposomal contents are based on azobenzene [60, 61]. Under ultraviolet irradiation (360 nm), azobenzene isomerizes from an extended *trans* form to a twisted *cis* form and returns back to the *trans* form under visible light (420 nm). Both steric effect and change in polarity resulting from this transformation can destabilize the vesicle. As the azobenzene groups undergo a wavelength-specific *cis–trans* isomerization, it is possible to achieve a pulsatile, externally controlled release. First phospholipids containing the azobenzene groups incorporated in fatty acyl chains were synthesized in 1990s. Recently the synthetic photosensitive lipids with azobenzene moiety in the structure are extensively studied. One example of such compounds is Bis-Azo PC (1,2-bis(4-n-butylphenylazo-4′-phenylbutyroyl)-L-α-phosphatidylcholine), which contains two acyl chains (see Fig. 5.4) [53] or $AzoC_{10}N^+$ (decyl-azobenzyl-triethylammonium) (see Fig. 5.5) [62].

Photoisomerizable liposomes could also be prepared using other photosensitive groups, such as spiropyran [63], spirooxazine [64] and retinoyl [65]. Liposomes with light-triggered release mechanism based on photoisomerization could have a significant advantage over standard liposomal formulations in the clinics due to their ability to reversibly release the drug or keep it inside the vesicle upon subsequent irradiation with appropriate wavelength.

5.4.1.3 Liposomes Triggered by Photopolymerization

Another mechanism of photo-triggering the release from liposomes is based on using lipids that can undergo photo-cross-linking leading to their photopolymerization within the liposome bilayer. In 1998, O'Brien and co-workers [66] reported that polymerization of bis-substituted lipids in the presence of other unreactive lipids resulted in the formation of domains. The same group showed that UV-induced cross-linking of lipids could effectively destabilize liposomes [54]. In their study they used UV-induced photopolymerization of bis-sorbyl phosphatidylcholine (bis-SorbPC) embedded in the liposomal bilayer in order to promote the release process. A different type of photo-polymerizable liposomes was obtained from the mixture of diacetylene phospholipid, 1,2-bis(tricosa-10,12-diynoyl)-sn-phosphatidylcholine ($DC_{8,9}PC$) and dipalmitoylphosphatidylcholine (DPPC) [55]. Photopolymerization of $DC_{8,9}PC$ was induced by 254 nm light. An interesting

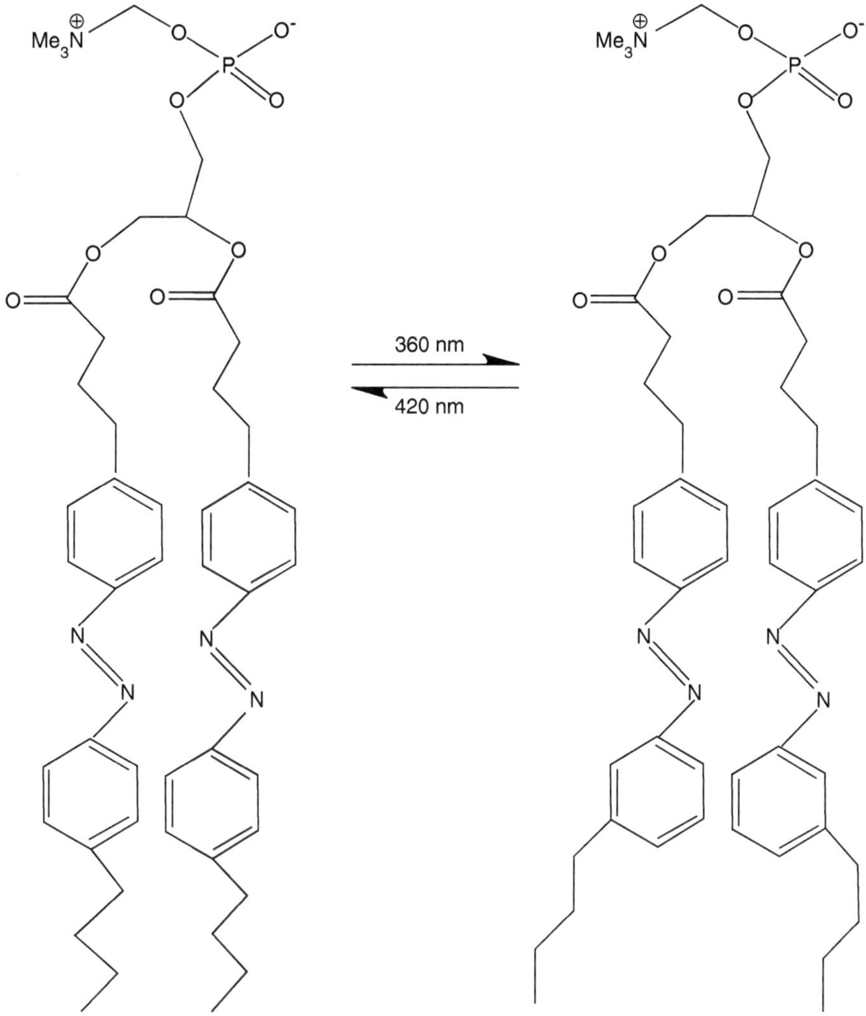

Fig. 5.4 The structure of Bis-Azo PC (1,2-bis(4-n-butylphenylazo-4′-phenylbutyroyl)-L-α-phosphatidyl-choline)

example of the practical application of photopolymerization-driven release was a liposomal system containing 10–20 mol% of $DC_{8,9}PC$, designed to release doxorubicin when illuminated with 514 nm laser light [67]. Other liposomal systems based on $DC_{8,9}PC$ were also developed [68].

Another mechanism utilized for inducing photo-triggered release from liposomes is based on photothermal processes. Photothermally induced release is driven by conversion of the absorbed light into heat with ensuing thermal or mechanical processes in the lipid membrane or the solute. The liposomal systems exhibiting the ability for the photothermally triggered release have been developed using various

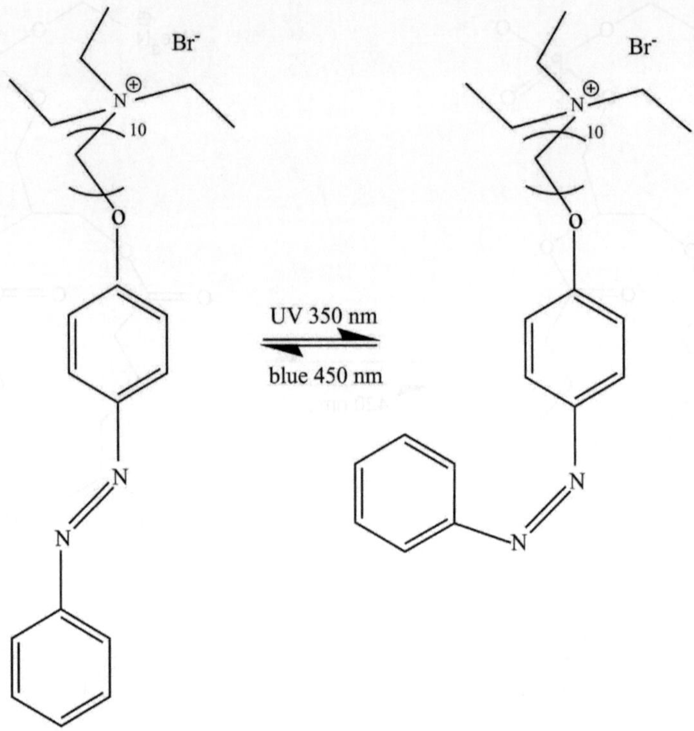

Fig. 5.5 The structure of AzoC$_{10}$N^{+} (decyl-azobenzyl-triethylammonium)

light-absorbing molecular dyes [69], such as calcein, sulforhodamine, indocyanine green (ICG) and 1,1′-dioctadecyl-3,3,3′,3′- tetramethylindodicarbocyanine iodide(DiD) (see Fig. 5.6).

5.4.2 Light-Responsive Polymeric Hydrogels

Already in the 1950s of the last century, stimuli-responsive hydrogels have been studied as a drug delivery system [70]. Light-sensitive hydrogels can change their physicochemical properties after exposition to light. Tanaka's group conducted interesting studies on the visible light-responsive hydrogels [71]. They introduced the trisodium salt of copper chlorophyllin into poly(N-isopropylacrylamide) hydrogels. When hydrogel is irradiated with appropriate wavelength, the chromophore absorbs the light, increasing the local temperature of the gel. The change of local temperature causes modification of the swelling behaviour of the gel. Application of this method is, however, limited due to the low two-photon absorption cross-sections of typical chromophores, which makes the process very inefficient, even when femtosecond pulse lasers are applied. Suzuki et al. [72] showed also that the same type of hydrogel may be enriched in other functional groups, thus gaining multi-responsive character, e.g. sensitivity to pH.

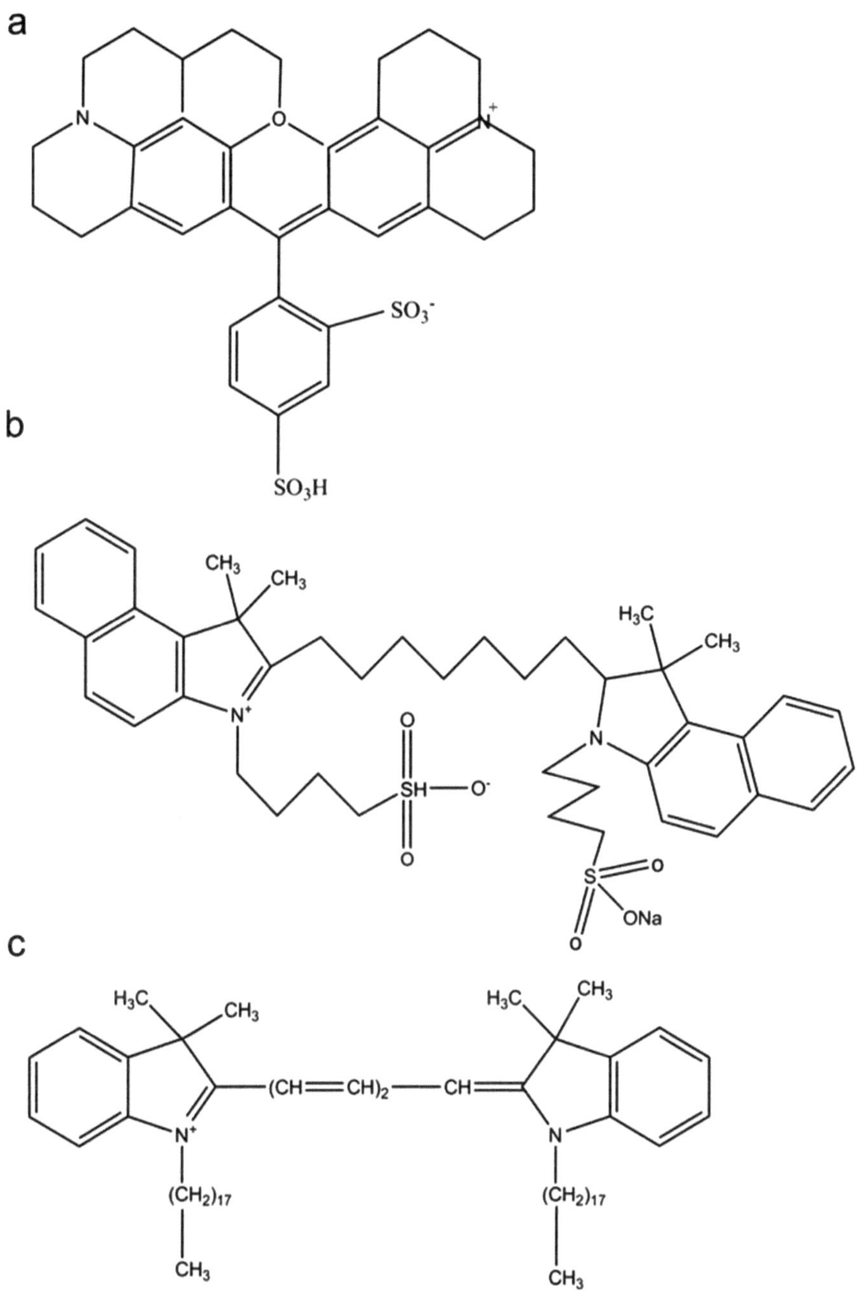

Fig. 5.6 The structure of the light-absorbing molecular dyes (**a**) sulforhodamine, (**b**) indocyanine green (ICG), (**c**) 1,1′-dioctadecyl-3,3,3′,3′- tetramethylindodicarbocyanine iodide(DiD)

Alternative strategy to design photoresponsive hydrogel systems is based on incorporating NIR-absorbing nanostructures, e.g. carbon nanotubes [73], gold nanoparticles [74] or graphene oxide [75], into thermosensitive hydrogel matrix. In all these instances the photothermal effect is used to convert near-infrared light into heat leading to changes in the hydrogel structure.

In 2012, Zhu and co-workers [75] presented an interesting example of a nanocomposite hydrogel system, where shape transitions of a photoresponsive hybrid material were achieved by exposure to near-infrared (NIR) laser. The system was based on poly(N-isopropylacrylamide) and graphene oxide (PNIPAM/GO) and was formed by introducing GO into the monomer solution of NIPAM before γ-irradiation-assisted polymerization. Graphene oxide has high optical absorbance; therefore the nanocomposite hydrogel shows great photothermal properties. The shape transitions of the material were reversible and could be easily controlled by on/off switching of the NIR laser irradiation. Biomacromolecules could be released from this system in aqueous media upon the sol–gel transition resulting from the irradiation with the NIR light source at 980 nm. The application of the hydrogels systems based on photothermal effect has two main constraints: (1) their construction requires thermosensitive polymers having a hydration–dehydration phase transition temperature and (2) the heating effect disappears after the NIR irradiation is turned off.

Upconversion nanoparticles (UCNPs) [76] are a new generation of fluorophores that can convert near-infrared radiation into visible radiation via a nonlinear optical process. In a study done by Yan's group, it was demonstrated that hydrogel loaded with UCNPs and irradiated by a CO_2 IR laser undergoes the gel–sol transition and can be used to release entrapped biomacromolecules [77]. The authors used the hydrogel matrix made from cross-linked hybrid polyacrylamide–poly(ethylene glycol) containing photo-cleavable o-nitrobenzyl moieties and core–shell lanthanide-doped UCNPs. The NIR-light-triggered degradation of a photosensitive hydrogel was caused directly by the UV light generated by encapsulated UCNPs. This work presented new, promising application of the unique multiphoton effect to obtain interesting photosensitive materials suitable for biomedical applications.

Another type of the hydrogels responds to light following the photodegradation mechanism. Disintegration of the polymer matrix upon irradiation allows for an external, real-time, spatial and temporal control over the hydrogel degradation and allows for staged and sequential release of a drug. Griffin and Kasko [78] described a series of macromeres containing the photodegradable ortho-nitrobenzyl (o-NB) linker in the macromere backbone. A series of hydrogels was prepared from these new macromeres via redox polymerization, and the apparent rate constants of degradation (k_{app}) of each gel were quantified using photo-rheology. Degradation rate could be controlled by introducing different functionalities into the o-NB linker structure. For example, decreasing the number of aryl ether substituents led to increased degradation constant, and the similar effect was observed upon changing the functionality at the benzylic site.

5.4.3 *Prodrugs*

In 1998 Wei et al. [79] reported the new model system of prodrug activation via photolysis. The authors used photolabile carbonate derivative of 5-fluorodeoxyuridine (5-FdU). Photolysis of 5-FdU derivative upon irradiation with long-wavelength UV light ($\lambda > 350$ nm) led to fast release of free 5-fluorodeoxyuridine. This carbonate ester of 5-FdU was potentially less toxic than unmodified 5-FdU, as it cannot undergo 5′-phosphorylation. Indeed the new prodrug was found to be non-toxic to cells but in the presence of UV irradiation ($\lambda = 350$ nm) inhibited the cell growth.

Another type of photo-triggered prodrugs is a system based on photocaging. Photocaging is an interesting method, in which a drug molecule is derivatized with a photolabile protecting group, so the molecule is biologically inert ("caged"). Light illumination of the prodrug results in the release of the caging group and activation of the drug. This strategy provides spatial, temporal and concentration-dependent control in drug delivery systems [80].

Another example of a prodrug system is based on double-activation prodrug system (drug—linker—deactivated photosensitizer), containing a photo-cleavable aminoacrylate linker and a deactivated photosensitizer. Hossion and co-workers [81] used three prodrugs containing either coumarin or one of the two potential drugs exhibiting high cytotoxicity (CA-4 and SN-38) to study both the activation of deactivated photosensitizer by cellular esterases and the release of the parent drugs upon irradiation with the visible light (540 nm) via photo-unclick chemistry. Furthermore, coumarin (a non-toxic prodrug) was used to quantify the release of parent drug in living cells. It was demonstrated that nearly 99% of coumarin was released from the prodrug after 24 h of incubation with MCF-7 cells followed by irradiation with low intensity visible light (8 mW/cm^2) for 30 min.

5.5 Conclusions

With all the many advantages of the light-triggered drug delivery systems, the most pronounced one being the ability to very precisely control the time, place and amount of released active agent, most of the described systems have not yet found their way to the clinics. This is mainly due to the still unsolved problem of limited distance that light can travel through the body tissue and resulting limitations in the treatment of solid tumours and large areas with pathological changes. Nevertheless, while it was primarily used almost exclusively on the skin, now light-triggered formulations are also applied in the therapies of gastrointestinal tract (light delivered through endoscope) and lungs (using bronchoscopy). What is more, with the further developments of the drug delivery systems, it should be possible to include in the light-triggered therapies in patients which could not be treated due to allergy to the free photoactive agents, e.g. porphyrins, or due to certain blood diseases.

References

1. M. Dobosz, S. Strobel, K.G. Stubenrauch, et al., Noninvasive measurement of pharmacokinetics by near-infrared fluorescence imaging in the eye of mice. J. Biomed. Opt. **19**(1), 016022 (2014)
2. A.M. Laughney, E. Kim, M.M. Sprachman, et al., Single-cell pharmacokinetic imaging reveals a therapeutic strategy to overcome drug resistance to the microtubule inhibitor eribulin. Sci. Transl. Med. **6**(261), 261ra152 (2014)
3. M.-H. Han, A.K. Friedman, Virogenetic and optogenetic mechanisms to define potential therapeutic targets in psychiatric disorders. Neuropharmacology **62**, 89–100 (2012)
4. A.A. Grace, S.B. Floresco, Y. Goto, et al., Regulation of firing of dopaminergic neurons and control of goal-directed behaviors. Trends Neurosci. **30**, 220–227 (2007)
5. J.L. Cao, H.E. Covington, A.K. Friedman, et al., Mesolimbic dopamine neurons in the brain reward circuit mediate susceptibility to social defeat and antidepressant action. J. Neurosci. **30**, 16453–16458 (2010)
6. C. Allen, D. Maysinger, A. Eisenberg, Nano-engineering block copolymer aggregates for drug delivery. Colloids Surf. B Biointerfaces **16**, 3–27 (1999)
7. C.-Y. Chen, C.-K. Syu, H.C. Lin, A stimulated mixed micelle system for in vitro study on chemo-photodynamic therapy. Macromol. Biosci. **16**, 188–197 (2016)
8. H.F. Wang, H.Z. Jia, S.X. Cheng, et al., PEG-stabilized micellar system with positively charged polyester core for fast pH-responsive drug release. Pharm. Res. **29**(6), 1582–1594 (2012)
9. J. Rios-Doria, A. Carie, T. Costich, et al., A versatile polymer micelle drug delivery system for encapsulation and in vivo stabilization of hydrophobic anticancer drugs. J. Drug. Deliv., 951741 (2012)
10. F. Sun, Y. Wang, Y. Wei, et al., Thermo-triggered drug delivery from polymeric micelles of poly(N-isopropylacrylamide-co-acrylamide)-b-poly(n-butyl methacrylate) for tumor targeting. J. Bioact. Compat. Polym. **29**(4), 301–317 (2014)
11. D.H. Thompson, O.V. Gerasimov, J.J. Wheeler, et al., Triggerable plasmalogen liposomes: Improvement of system efficiency. Biochim. Biophys. Acta **1279**, 25–34 (1996)
12. Y. Li, K. Xiao, W. Zhuet, et al., Stimuli-responsive cross-linked micelles for on-demand drug delivery against cancers. Adv. Drug Deliv. Rev. **66**, 58–73 (2014)
13. H. Yin, L. Liao, J. Fang, Enhanced permeability and retention (EPR) effect based tumor targeting: The concept, application and Prospect. JSM Clin. Oncol. Res. **2**(1), 1010 (2014)
14. A.M. Master, M. Livingston, N.L. Oleinick, et al., Optimization of a Nanomedicine-based silicon Phthalocyanine 4 photodynamic therapy (pc 4-PDT) strategy for targeted treatment of EGFR-overexpressing cancers. Mol. Pharm. **9**(8), 2331–2338 (2012)
15. C. Conte, S. Maiolino, D.S. Pellosi, et al., Polymeric nanoparticles for cancer photodynamic therapy. Top. Curr. Chem. **370**, 61–112 (2016)
16. C.S. De Paula, A.C. Tedesco, F.L. Primoet, et al., Chloroaluminium phthalocyanine polymeric nanoparticles as photosensitisers: Photophysical and physicochemical characterisation, release and phototoxicity in vitro. Eur. J. Pharm. Sci. **49**(3), 371–381 (2013)
17. P. Nalawade, B. Aware, V.J. Kadam, et al., Layered double hydroxides: A review. J. Sci. Ind. Res. (India) **68**, 267–272 (2009)
18. R.K. Kankala, Y. Kuthati, S.H.W. al, Multi-laminated metal hydroxide nanocontainers for oral-specific delivery for bioavailability improvement and treatment of inflammatory paw edema in mice. J. Colloid Interface Sci. **458**, 217–228 (2015)
19. K. Khorsandi, R. Hosseinzadeh, M. Fateh, Curcumin intercalated layered double hydroxide nanohybrid as a potential drug delivery system for effective photodynamic therapy in human breast cancer cells. RSC Adv. **5**, 93987–93994 (2015)
20. P.-R. Wei, Y. Kuthati, R.K. Kankala, et al., Synthesis and characterization of chitosan-coated near-infrared (NIR) layered double hydroxide-Indocyanine green Nanocomposites for potential applications in photodynamic therapy. Int. J. Mol. Sci. **16**(9), 20943–20968 (2015)

21. M. Merchan, T.S. Ouk, P. Kubat, et al., Photostability and photobactericidal properties of porphyrin-layered double hydroxide–polyurethane composite films. J. Mater. Chem. B **1**, 2139–2146 (2013)
22. X.-S. Li, M.-R. Ke, W. Huang, et al., A pH-responsive layered double hydroxide (LDH)–Phthalocyanine nanohybrid for efficient photodynamic therapy. Chemistry **21**, 3310–3317 (2015)
23. B.G. Trewyn, I.I. Slowing, S. Giri, et al., Synthesis and functionalization of a Mesoporous silica nanoparticle based on the sol–gel process and applications in controlled release. Acc. Chem. Res. **40**, 846–853 (2007)
24. M. Arduini, F. Mancin, P. Tecilla, et al., Self-organized fluorescent nanosensors for ratiometric Pb2þ detection. Langmuir **23**, 8632–8636 (2007)
25. I. Roy, T.Y. Ohulchanskyy, H.E. Pudavaret, et al., Ceramic-based nanoparticles entrapping water-insoluble photosensitizing anticancer drugs: A novel drug−carrier system for photodynamic therapy. J. Am. Chem. Soc. **125**, 7860–7865 (2003)
26. F. Selvestrel, F. Moret, D. Segat, et al., Targeted delivery of photosensitizers: Efficacy and selectivity issues revealed by multifunctional ORMOSIL nanovectors in cellular systems. Nanoscale **5**, 6106–6116 (2013)
27. D.J. Evans, The bionanoscience of plant viruses: templates and synthons for new materials. J. Mater. Chem. **18**, 3746–3754 (2008)
28. Q. Zeng, S. Saha, L.A. Lee, et al., Chemoselective modification of turnip yellow mosaic virus by cu(I) catalyzed azide−alkyne 1,3-dipolar cycloaddition reaction and its application in cell binding. Bioconjug. Chem. **22**, 58–66 (2011)
29. B.A. Cohen, M. Bergkvist, Targeted in vitro photodynamic therapy via aptamer-labeled, porphyrin-loaded virus capsids. J. Photochem. Photobiol. B **121**, 67–74 (2013)
30. A.G. Havanessiean, C. Soundaramourty, D. El Khoury, et al., Surface expressed nucleolin is constantly induced in tumor cells to mediate calcium-dependent ligand internalization. PLoS One **5**, e15787 (2010)
31. D. Gabriel, M.F. Zuluaga, N. Lange, On the cutting edge: protease-sensitive prodrugs for the delivery of photoactive compounds. Photochem. Photobiol. Sci. **10**, 689–703 (2011)
32. R. Weissleder, C.H. Tung, U. Mahmood, et al., In vivo imaging of tumors with protease-activated near-infrared fluorescent probes. Nat. Biotechnol. **17**(4), 375–378 (1999)
33. E. Gounaris, C.H. Tung, C. Restaino, et al., Live imaging of cysteine-cathepsin activity reveals dynamics of focal inflammation, angiogenesis, and polyp growth. PLoS One **3**, e2916 (2008)
34. M.R. Hamblin, J.L. Miller, I. Rizvi, et al., Pegylation of charged polymer-photosensitiser conjugates: effects on photodynamic efficacy. Br. J. Cancer **89**, 937–943 (2003)
35. Y. Choi, R. Weissleder, C.H. Tung, Selective antitumor effect of novel protease-mediated photodynamic agent. Cancer Res. **66**, 7225–7229 (2006)
36. B.C. Bae, K. Na, Self-quenching polysaccharide-based nanogels of pullulan/folate-photosensitizer conjugates for photodynamic therapy. Biomaterials **31**, 6325–6335 (2010)
37. F. Caruso, M. Rossi, A. Benson, et al., Ruthenium–Arene complexes of curcumin: X-ray and density functional theory structure, synthesis, and spectroscopic characterization, in vitro antitumor activity, and DNA docking studies of (p-cymene)Ru(curcuminato)chloro. J. Med. Chem. **55**, 1072–1081 (2012)
38. D. Pucci, A. Crispini, B.S. Mendiguchía, et al., Improving the bioactivity of Zn(II)-curcumin based complexes. Dalton Trans. **42**, 9679–9687 (2013)
39. A.K. Renfrew, N.S. Bryce, T.W. Hambley, Delivery and release of curcumin by a hypoxia-activated cobalt chaperone: A XANES and FLIM study. Chem. Sci. **4**, 3731–3739 (2013)
40. R. Pettinari, F. Marchetti, F. Condello, et al., Ruthenium(II)-Arene RAPTA type complexes containing curcumin and bisdemethoxycurcumin display potent and selective anticancer activity. Organometallics **33**, 3709−3715 (2014)
41. T.K. Goswami, S. Gadadharb, B. Gole, et al., Photocytotoxicity of copper(II) complexes of curcumin and N-ferrocenylmethyl-l-amino acids. Eur. J. Med. Chem. **63**, 800−810 (2013)
42. A. Hussain, K. Somyajit, B. Banik, et al., Enhancing the photocytotoxic potential of curcumin on terpyridyl lanthanide(III) complex formation. Dalton Trans. **42**, 182−195 (2013)

43. S. Banerjee, I. Pant, I. Khan, Remarkable enhancement in photocytotoxicity and hydrolytic stability of curcumin on binding to an oxovanadium(IV) moiety. Dalton Trans. **44**, 4108–4122 (2015)
44. S. Banerjee, P. Prasad, I. Khan, et al., Mitochondria targeting photocytotoxic oxidovanadium (IV) complexes of curcumin and (acridinyl) dipyridophenazine in visible light. Anorg. Allg. Chem. **640**, 1195–1204 (2014)
45. A.W. Carpenter, M.H. Schoenfisch, Nitric oxide release: Part II. Therapeutic applications. Chem. Soc. Rev. **41**, 3742 (2012)
46. L.K. Keefer, Nitric oxide (NO)- and Nitroxyl (HNO)-generating diazeniumdiolates (NONOates): emerging commercial opportunities. Curr. Top. Med. Chem. **5**, 625–636 (2005)
47. B. Heilman, P.K. Mascharak, Light-triggered nitric oxide delivery to malignant sites and infection. Phil. Trans. R. Soc. A **371**, 20120368 (2013)
48. B.J. Heilman, G.M. Halpenny, P.K. Mascharak, Synthesis, characterization, and light-controlled antibiotic application of a composite material derived from polyurethane and silica xerogel with embedded photoactive manganese nitrosyl. Biomed. Mater. Res. B **99B**, 328–337 (2011)
49. J. Xu, F. Zeng, H. Wu, et al., A mitochondrial-targeting and NO-based anticancer nanosystem with enhanced photo-controllability and low dark-toxicity. J. Mater. Chem. B **3**, 4904–4912 (2015)
50. B. Chandra, R. Subramaniam, S. Mallik, et al., Formulation of photocleavable liposomes and the mechanism of their content release. Org. Biomol. Chem. **4**, 1730–1740 (2006)
51. A. Pashkovskaya, E. Kotova, Y. Zorlu, et al., Light-triggered liposomal release: membrane permeabilization by photodynamic action. Langmuir **26**, 5726–5733 (2010)
52. Z.Y. Zhang, B.D. Smith, Synthesis and characterization of NVOC-DOPE, a caged photoactivatable derivative of Dioleoylphosphatidylethanolamine. Bioconjug. Chem. **10**, 1150–1152 (1999)
53. Morgan CG, , Bisby RH, Johnson SA et al (1995) Fast solute release from photosensitive liposomes: an alternative to 'caged' reagents for use in biological systems. FEBS Lett. 375:113–116
54. B. Bondurant, D.F. O'Brien, Photoinduced destabilization of Sterically stabilized liposomes. J. Am. Chem. Soc. **120**, 13541–13542 (1998)
55. A. Yavlovich, A. Singh, S. Tarasov, et al., Design of liposomes containing photopolymerizable phospholipids for triggered release of contents. J. Therm. Anal. Calorim. **98**, 97–104 (2009)
56. B. Khoobehi, G.A. Peyman, N. Bhatt, et al., Laser-induced experimental vascular occlusion using liposome-encapsulated ADP. Lasers Surg. Med. **12**, 609–614 (1992)
57. S. Mordon, T. Desmettre, J.-M. Devoisselle, et al., Thermal damage assessment of blood vessels in a hamster skin flap model by fluorescence measurement of a liposome-dye system. Lasers Surg. Med. **20**, 131–141 (1997)
58. D.L. Van der Meullen, P. Misra, J. Michael, et al., Laser mediated release of dye from liposomes. Photochem. Photobiol. **56**, 325–332 (1992)
59. Z. Li, Y. Wan, A.G. Kutateladze, Dithiane-based photolabile amphiphiles: toward photolabile liposomes. Langmuir **19**, 6381–6391 (2003)
60. K. Kano, Y. Tanaka, T. Ogawa, et al., Photoresponsive artificial membrane. Regulation of membrane permeability of liposomal membrane by photoreversible Cis-trans isomerization of azobenzenes. Photochem. Photobiol. **34**, 323–329 (1981)
61. T. Sato, T. Phoeung, P.-A. Rousseau, et al., Nonphospholipid fluid liposomes with switchable photocontrolled release. Langmuir **7**, 2330–2335 (1991)
62. Z.K. Cui, Y. Phoeung, P.A. Rousseau, et al., Nonphospholipid fluid liposomes with switchable photocontrolled release. Langmuir **30**, 10818–10825 (2014)
63. R.F. Khairutdinov, J.K. Hurst, Photocontrol of ion permeation through bilayer membranes using an amphiphilic spiropyran. Langmuir **17**, 6881–6886 (2001)
64. R.F. Khairutdinov, K. Giertz, J.K. Hurst, et al., Photochromism of spirooxazines in homogeneous solution and phospholipid liposomes. J. Am. Chem. Soc. **120**, 12707–12713 (1998)
65. C. Pidgeon, C.A. Hunt, Light sensitive liposomes. Photochem. Photobiol. **37**, 491–494 (1983)
66. D.F. O'Brien, B. Armitage, A. Benedicto, et al., Polymerization of preformed self-organized assemblies. Acc. Chem. Res. **31**, 861–868 (1998)

67. A. Yavlovich, B. Smith, K. Gupta, et al., Light-sensitive lipid-based nanoparticles for drug delivery: design principles and future considerations for biological applications. Mol. Membr. Biol. **27**(7), 364–381 (2010)
68. A. Yavlovich, A. Singh, R. Blumenthal, et al., A novel class of photo-triggerable liposomes containing DPPC:DC$_{8,9}$PC as vehicles for delivery of doxorubcin to cells. Biochim. Biophys. Acta **1808**, 117–126 (2011)
69. S.J. Leung, M. Romanowski, Light-activated content release from liposomes. Theranostics **2**(10), 1020–1036 (2012)
70. W. Kuhn, B. Hargitay, A. Katchalsky, et al., Reversible dilation and contraction by changing the state of ionization of high-polymer acid networks. Nature **165**, 514–516 (1950)
71. A. Suzuki, T. Tanaka, Phase transition in polymer gels induced by visible light. Nature **346**, 345–347 (1990)
72. A. Suzuki, T. Ishii, Y. Maruyama, Optical switching in polymer gels. J. Appl. Phys. **80**(1), 131–136 (1996)
73. S.K. Samanta, A. Pal, S. Bhattacharya, et al., Carbon nanotube reinforced supramolecular gels with electrically conducting, viscoelastic and near-infrared sensitive properties. J. Mater. Chem. **20**, 6881–6890 (2010)
74. A. Shiotani, T. Mori, T. Niidome, et al., Stable incorporation of gold Nanorods into N-Isopropylacrylamide hydrogels and their rapid shrinkage induced by near-infrared laser irradiation. Langmuir **23**, 4012–4018 (2007)
75. C.-H. Zhu, Y. Lu, J. Peng, et al., Photothermally sensitive poly(N-isopropylacrylamide)/graphene oxide Nanocomposite hydrogels as remote light-controlled liquid microvalves. Adv. Funct. Mater. **22**(19), 4017–4022 (2012)
76. M. Wang, G. Abbineni, A. Clevenger, et al., Upconversion nanoparticles: synthesis, surface modification and biological applications. Nanomedicine **7**(6), 710–729 (2011)
77. B. Yan, J.-C. Boyer, D. Habault, et al., Near infrared light triggered release of biomacromolecules from hydrogels loaded with upconversion nanoparticles. J. Am. Chem. Soc. **134**(40), 16558–16561 (2012)
78. D.R. Griffin, A.M. Kasko, Photodegradable macromers and hydrogels for live cell encapsulation and release. J. Am. Chem. Soc. **134**(31), 13103–13107 (2012)
79. Y. Wei, Y. Yan, D. Pei, et al., A photoactivated prodrug. Bioorg. Med. Chem. Lett. **8**, 2419–2422 (1998)
80. T. Joshi, V. Pierroz, C. Mari, et al., A bis (dipyridophenazine) (2-(2-pyridyl)pyrimidine-4-carboxylicacid) ruthenium(II) complex with anticancer action upon photodeprotection. Angew. Chem. Int. Ed. **53**, 2960–2963 (2014)
81. A.M.L. Hossion, M. Bio, G. Nkepang, et al., Visible light controlled release of anticancer drug through double activation of prodrug. ACS Med. Chem. Lett. **4**(1), 124–127 (2013)

Chapter 6
Gene Therapy Approaches Toward Biomedical Breakthroughs

Lies Marguillier, Peter Dubruel, and Sandra Van Vlierberghe

6.1 Introduction

In general, the expression of a protein is the outcome of a complicated, highly controlled multistep process in cells (see Fig. 6.1), which is often referred to as "central dogma" [5]. Most genes contain information enabling protein synthesis and some produce other molecules that help the cell to assemble these proteins. The process of gene expression consists of two major steps: transcription and translation. During the transcription step, the information stored in DNA consisting of a series of nucleotides is transferred into a similar molecule (i.e., RNA), in the cell nucleus. The type of RNA that carries the information from the DNA out of the nucleus into the cytoplasm is defined as messenger RNA (mRNA). The translation step occurs in the cytoplasm. There, the mRNA interacts with a ribosome, which reads the sequence of mRNA bases. Each sequence of three bases (aka codon) codes for one particular amino acid. A second type of RNA, namely, transfer RNA (tRNA), assembles the protein by adding amino acids one by one to the protein chain. As soon as the ribosome encounters a stop codon, the protein assembly process finishes.

L. Marguillier
Polymer Chemistry & Biomaterials Group, Ghent University, Ghent, Belgium

P. Dubruel
Polymer Chemistry & Biomaterials Group, Centre of Macromolecular Chemistry, Department of Organic and Macromolecular Chemistry, Ghent University, Ghent, Belgium

S. Van Vlierberghe (✉)
Polymer Chemistry & Biomaterials Group, Centre of Macromolecular Chemistry, Department of Organic and Macromolecular Chemistry, Ghent University, Ghent, Belgium

Brussels Photonics, Department of Applied Physics and Photonics, Vrije Universiteit Brussel, Brussel, Belgium
e-mail: sandra.vanvlierberghe@ugent.be

J. Van Hoorick et al. (eds.), *Polymer and Photonic Materials Towards Biomedical Breakthroughs*, Micro- and Opto-Electronic Materials, Structures, and Systems, https://doi.org/10.1007/978-3-319-75801-5_6

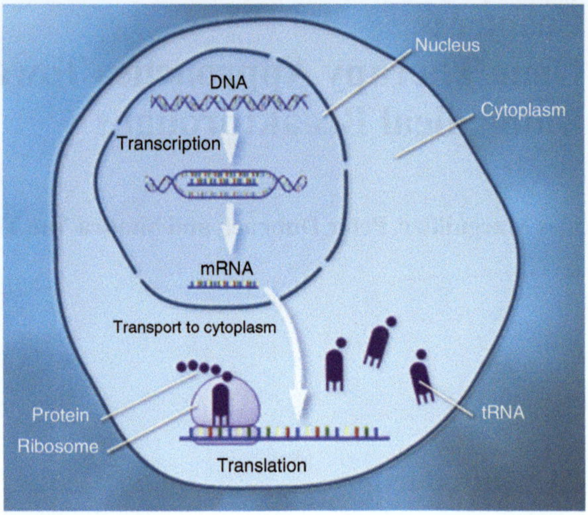

Fig. 6.1 Illustration of the central dogma: information from genes is applied to produce proteins through the processes of transcription and translation [3]

A mutation occurring in a gene can result in the faulty production of a protein or even in a total lack of protein synthesis. These phenomena can hamper the correct functioning of cells [6]. A possible solution to this problem is offered by gene therapy, which aims to treat the disease at a cellular and at a molecular level. Gene therapy addresses the root cause of diseases, which is in great contrast with the current symptomatic treatments.

Free oligonucleotides and DNA are rapidly degraded by serum nucleases in the blood when injected intravenously. Furthermore, the delivery of naked DNA in a cell is very inefficient since DNA molecules have very large sizes and are highly negatively charged and therefore electrostatically repulsed by the negatively charged cell membranes. These problems result in a very low transfer efficiency of naked DNA and thereby indicate the need for the design of effective carriers that compact and protect oligonucleotides [7]. In gene therapy, these carriers are often referred to as vectors.

Methods to deliver genes or plasmid DNA to cells by means of a vector can be divided in two classes. The first is viral gene delivery, which employs genetically altered viruses that have their genome stripped in order to prevent viral replication, to reduce cytotoxicity, and to permit incorporation of the therapeutic transgene. The process by which the foreign DNA is introduced into the cells via a viral vector is called transduction. The second class of delivery systems is referred to as non-viral gene delivery. It involves the use of DNA either as such or in combination with synthetic carrier molecules including cationic lipids or polymers. The non-viral introduction of DNA into a cell and its subsequent integration into the cell's chromosomal DNA is referred to as transfection.

Two different approaches exist to introduce DNA combined with a viral or non-viral vector in a cell [8]. Both of them are represented in Fig. 6.2. The first approach is the in vivo approach, in which genes are directly delivered to the living organism. The latter can be performed by direct infusion or injection of the DNA-vector complex or by implantation of a gene-activated matrix [9]. All of these methods result in the transfection or transduction of the cells in situ by the delivery and release of the gene inside the patient [2]. The second approach is called ex vivo and requires the

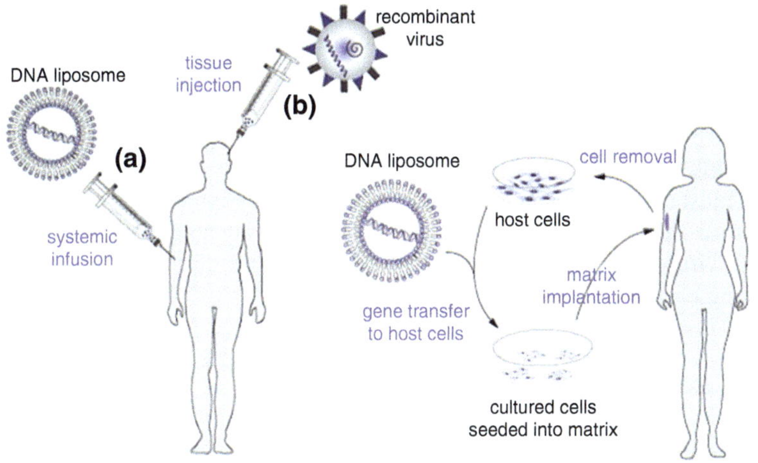

Fig. 6.2 Graphical representation of **in vivo** gene therapy including systemic infusion of DNA liposomes and tissue injection of recombinant viruses (**left**) and **ex vivo** gene therapy using DNA liposomes (**right**)

explantation of cells from the host. The gene of interest can then be transferred to the targeted cells using a vector. In a subsequent step, the cells are grown in culture to be transfected or transduced in vitro. This procedure is followed by cell seeding on a suitable matrix. The resultant cell supplemented matrix can then finally be implanted into the host tissue. Advantages of the latter approach include its safety and effectiveness, while the former approach is attractive because of its technical simplicity. However, the disadvantages of the in vivo approach are that the targeting is not completely specific and that the transfection efficiency is limited [10]. The main drawback of the ex vivo approach is the complexity of the human body that requires appropriate cell sourcing and intensive surgical procedures to achieve targeted delivery [11].

6.2 Overview of Various Gene Delivery Systems

6.2.1 Viral Gene Delivery

Viruses are ultramicroscopic organisms that bind to their host and introduce their genetic material into the host cells as part of their replication cycle. The host cell carries out the instructions present in the viral genes and produces additional proteins and copies of the virus. This leads to an increasing amount of cells that become infected with the genetic material, thereby rapidly proliferating at the expense of the host cell. As a result, viruses are obvious candidates to deliver transgenes to cells.

Initial research in gene therapy thus focused on the use of viral carriers to compact and protect oligonucleotides. Viruses are genetically altered in order to prevent viral replication, to reduce cytotoxicity, and to permit incorporation of the therapeutic transgene. The main advantage of viral vectors is that they exhibit a high efficiency to deliver both DNA and RNA to numerous cell lines. They are extremely efficient in their expression, with essentially only one single viral particle being required to induce measurable effects [7]. Disadvantages of viral vector systems include their DNA capacity, cytotoxicity, possible immunogenic responses, and limitations with respect to scaling-up procedures. In addition, some viral vectors are incapable of transfecting nondividing cells [12].

The most commonly applied viruses in gene therapy include the first-generation retroviruses and adenoviruses and the more recent adeno-associated and herpes simplex viruses. All of the abovementioned types are discussed in detail in the upcoming paragraphs.

6.2.1.1 Retroviruses

Retroviruses are most often applied in clinical trials investigating viral vectors for gene delivery. The most frequently used retrovirus is derived from the murine leukemia virus [1] (i.e., an enveloped onco-retrovirus with a genome of linear, single-stranded RNA (ss-RNA)).

Interestingly, retroviruses selectively infect proliferating cells. The genetic material in retroviruses is present in the form of RNA molecules, while their host cell's genetic material is in the form of DNA. A retrovirus infecting a host cell introduces its RNA in combination with several enzymes into the cell. A DNA copy of this RNA should be produced to enable integration of the viral RNA in the host cell. The copying process is referred to as reverse transcription and is carried out by one of the viral enzymes (reverse transcriptase) [13]. Next, the DNA copy resides within the nucleus of the host cell and should be incorporated into the host cell's genome. A second enzyme, called integrase, regulates the insertion of the DNA copy into the chromosomes [14]. Once the viral genetic material has been inserted, all the future descendants of the host cell will contain the new genes.

The major advantage of retroviral vectors is their integration into the host cell chromosome, offering stable transformation and sustained gene expression. However, through non-specific integration into the host chromosome, there exists a certain risk of insertional mutagenesis [10]. Other disadvantages include a limited transduction efficiency and a rather small DNA size (up to 8.5 kb) [4].

6.2.1.2 Adenoviruses

In contrast with retroviruses, adenoviruses can infect both dividing and nondividing cells. They enter their host cells via receptor-mediated pathways, and the viral genome is not integrated into the host genome. Adenoviruses have the ability to infect a broad range of cells including the skin, lung, liver, brain, muscle, and blood vessels [10]. They are efficient vectors, since a very high percentage of infected cells expresses the transgene. All of these features render adenoviruses promising and versatile vectors for gene therapy.

Adenoviruses carry their genetic material in the form of double-stranded DNA (ds-DNA). When infecting a host cell, they introduce their DNA, which is released

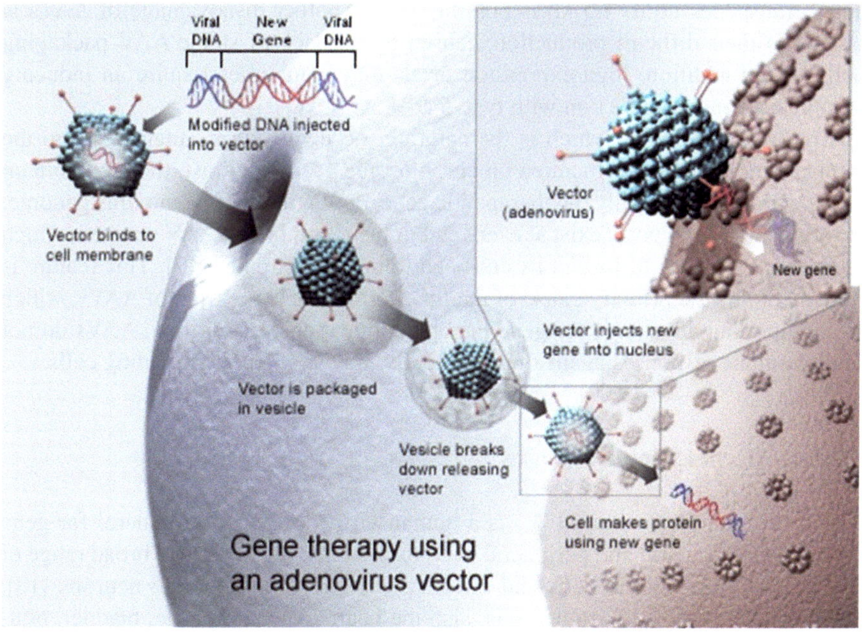

Fig. 6.3 Schematic representation of the transduction mechanism of adenoviruses: a new gene is inserted into the adenovirus vector applied to introduce the modified DNA into a human cell [3]

in the cell nucleus. The process is schematically represented in Fig. 6.3. The instructions in the viral DNA are transcribed and translated in an identical way as for any other gene in the host cell. Contrary to retroviruses, the introduced DNA is not replicated as the cell divides, implying that the descendants of that cell will not contain the viral DNA [15]. Hence, treatments with adenoviruses require readministration in a growing cell population.

Adenoviruses show also several drawbacks that limit their suitability for gene therapy applications. The major limitations of an adenovirus include the antigenicity that the virus imparts to transduced cells and the immunogenicity of the virus particle as such [4]. As a result, transgene expression is often limited to short time periods [10]. Other drawbacks include that the viral DNA is not transmitted to all daughter cells upon cell division. In addition, although adenoviruses infect already many cell types, at present not all cell types can be infected.

6.2.1.3 Adeno-Associated Viruses

Adeno-associated viruses (AAVs) are small viruses that belong to the parvovirus family. Their genome consists of ss-DNA, and since they are naturally replication-deficient, they can only proliferate in the presence of helper viruses such as adenoviruses or herpes viruses. Because of their limited genome size, only insertion of small transgenes up to 4.5 kb is possible [10]. Another disadvantage of AAVs is related to their difficult production, caused by the lack of stable AAV packaging cells [4]. In addition, high expression levels may sometimes require an inducing stimulus, such as coinfection with type 5 adenoviruses [16].

The wild type AAV, which is the naturally occurring form, integrates into the human genome at a specific area on chromosome 19. In contrast, the recombinant AAV, which contains only the therapeutic gene, does not integrate into this genome. Inverted terminal repeats exist at every end of the recombinant AAV genome, which enables the genome to fuse at its ends and to become circular [17]. This feature is assumed to be the primary cause of the long-term gene expression of AAVs, which is an important advantage for gene therapy applications. In addition, AAVs do not elicit immune responses against infected cells and can infect nondividing cells.

6.2.1.4 Herpes Simplex Viruses

The herpes simplex virus (HSV) is a human virus developed in general for gene delivery to the central and peripheral nervous systems. It can affect a broad range of mitotic and post-mitotic cells and effectively deliver transgenes to neurons [18]. Interestingly, in addition to neurons, also the heart, skeletal muscle, bladder, pituitary gland, articular joint, peripheral blood, inguinal adipose tissue, and mononuclear cells can be infected by HSVs [10]. HSV vectors have been investigated extensively to treat central nervous system tumors and other degenerative diseases.

HSV is a very large virus that can accommodate transgenes far larger (up to 50 kb) than any other viral vector. This extended carrying capacity is a potential advantage of the virus. Furthermore, HSV has a remarkable ability to persist in infected cells and to evade the immune system due to its latency. This latency is associated with the silencing of almost all viral transcription phenomena [4]. However, a possible drawback is the fact that transgene expression also becomes quiescent during latency. For this reason HSVs offer a great promise mainly for short-term gene delivery applications [19].

The first-generation vectors (i.e., retroviruses and adenoviruses) enabled gene transfer but showed difficulties toward implementation into clinical practice. However, more recently AAVs and HSVs showed some substantial improvements although they are still associated with several fundamental, immunogenic problems. Interestingly, non-viral gene delivery is recently emerging as a realistic alternative for the use of viral vectors.

6.2.2 *Non-viral Gene Delivery*

Non-viral gene delivery involves the use of therapeutic DNA either as such (i.e., naked DNA) or complexed with carrier molecules including cationic lipids or polymers. The main advantages of non-viral gene delivery systems are their biocompatibility, their limited risk of insertion mutagenesis, their potential enabling large-scale

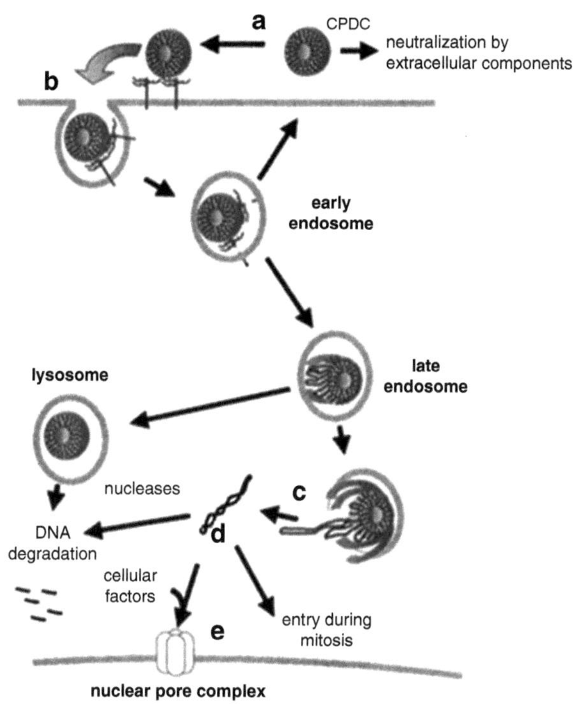

Fig. 6.4 Schematic representation of non-viral gene delivery barriers for cationic polymer/DNA complexes (CPDCs). These barriers include the stability of the DNA in the complex (**a**), the cellular uptake by endocytosis (**b**), the escape from the endosomal compartments (**c**), the cytosolic transport to the nucleus (**d**), and the entry into the nucleus (**e**) [12]

production, and their lower cost [20]. In addition, non-viral systems offer a better control of their molecular composition and a higher flexibility in the size of the transgenes to be delivered [12]. Moreover, non-viral gene transfer is usually characterized by transient gene expression. In clinical settings related to wound healing or bone regeneration, this short-term expression may be desirable [10].

Unfortunately, up to now non-viral gene carriers still exhibit a significantly reduced transfection efficiency compared to viral carriers. The non-viral analogues generally show an inability to overcome the numerous extra- and intracellular physical and chemical boundaries between the site of administration and the cell nucleus [7]. As illustrated in Fig. 6.4, this series of barriers includes the physical and chemical stability of DNA, the cellular uptake by endocytosis, the escape from the endosomal compartments prior to encountering lysosomes, the cytosolic transport, and the entry into the nucleus [12]. The abovementioned barriers are discussed in depth in section "Cationic Polymers". In addition to these barriers, biological limitations are also present. The main biological barriers include the immunogenic response to the vector as such and the immune stimulation by specific DNA sequences. However, by optimizing the vector and the plasmid sequence, biological problems can be minimized, rendering the physical and chemical barriers the major issues determining the transfection efficiency.

As a result of the attractive properties of non-viral vector systems, various researchers have been focusing in the past decade on the design of compounds that can form complexes with DNA and which can avoid both the in vitro and in vivo barriers for gene delivery [7]. Various non-viral methods exist to introduce foreign DNA into the targeted cells. Some rely on physical processes, while others rely on chemical materials. In the upcoming sections, various relevant transfection methods are described.

6.2.2.1 Physical Transfection Methods

A first physical transfection method is the direct injection of naked DNA. It is known as the simplest and safest method to enable gene delivery. Important advantages include the unlimited gene size to be delivered and the lack of any toxicity whatsoever. Via direct injection using hypodermic needles, successful delivery and gene expression have been achieved in different tissues [21]. Limitations of this transfection method include its low efficiency and the occurrence of transient gene expression. Furthermore, the method is only applicable to directly accessible tissue [10]. Direct DNA injection is shown to be most efficient in cardiac and skeletal muscle cells.

Extensive research has been performed regarding more efficient methods for the delivery of naked DNA. A first interesting method is electroporation. Using this method, brief electrical pulses are applied to cells to transiently create pores in the plasma membrane, allowing DNA diffusion into the cell cytoplasm. Pore formation occurs by exposing cells to intense electrical fields inducing a transmembrane potential. When this potential possesses sufficient magnitude, it triggers a dielectric breakdown of the cell membrane, which results in pore formation [22]. Similar to the gene gun technique, electroporation is advantageous because of its technical simplicity and the fact that large DNA amounts can be delivered. Disadvantages include the non-specificity, the cell membrane damage required for DNA uptake, and the need for complex equipment [10].

A second technique to enhance efficient delivery of DNA is particle bombardment, which is also referred to as gene gun. Using this approach, DNA is coupled to a nanoparticle of an inert solid, often gold. The resulting DNA-coated gold particles are then accelerated to be shot directly into the target cell's nucleus [23]. This acceleration is usually induced by means of high-pressure helium gas. The gene gun approach is technically straightforward and has the advantage that large DNA amounts can be delivered. Important drawbacks are its non-specificity, the introduction of foreign particles (i.e., gold particles) into the tissue, and the physical damage (cfr. membrane puncture) to the targeted cells [10].

In addition to the abovementioned techniques, alternative physical transfection methods also exist, including impalefection, sonoporation, optical transfection, and magnet-assisted transfection [24]. These techniques are less frequently applied for gene therapy applications.

6.2.2.2 Chemical Transfection Methods

Cationic Polymers

Different chemical transfection methods exist, with cationic polymers being the main topic of the present work. Therefore, their cellular uptake mechanism will be described in detail. Nevertheless, various other gene delivery processes are anticipated to occur in a similar way. Cationic polymers can be applied to enhance gene transfection by protecting DNA molecules from nuclease degradation and enabling their passage through the cell membrane [25]. Since the polymers are positively charged, they have a natural affinity for the negatively charged DNA and for the cell membrane [26]. The latter is of relevance for their transfection mechanism.

Transfection Mechanism

As already highlighted before (Sect. 6.2.2), the limited transfection efficiency of non-viral vectors is primarily due to the presence of several barriers between the site of administration and the delivery to the cell nucleus. Cationic polymers can be applied with the aim to offer strategies to overcome these obstacles [7]. The first important concept is DNA complexation. Facile cellular uptake of free DNA via plasma membrane permeation is hindered by the size and by the negative charge of DNA. DNA complexation with cationic polymers is possible due to electrostatic interactions. The complex formation results in charge neutralization and DNA compaction and therefore facilitates cellular uptake. Polymer-DNA complexes are often referred to as polyplexes (see Sect. 1.1 of Chap. 3). The stability of the delivery system as such is related both to the chemical stability of the DNA and the physical stability of the delivery vector [12].

The formed complexes can be divided into two groups, based on whether or not they are conjugated to targeting ligands. For the nontargeted cationic complexes, research suggests that the complexes first associate with the cell membrane through electrostatic interactions with anionic heparan sulfate proteoglycans (HSPGs), present on the cell surface. These interactions are modulated by the overall complex charge [12]. After this association, cellular uptake of the cationic complexes can proceed through various endocytotic mechanisms. Endocytosis is defined as the process in which substances from the extracellular environment are internalized through the formation of vesicles formed by membrane fusion. The resulting vesicles are referred to as endosomes. The second group of cationic complexes, containing vector systems with attached receptor ligands to promote cell- and tissue-specific delivery, exhibits a similar transfection route. The main difference between the two groups is the effect of complex size on the cellular uptake. While for nontargeted cationic polyplexes large particles have exhibited more successful gene delivery in some cases, this was not the case for targeted cationic complexes [7].

A very important step in the series of boundaries that the delivery systems have to overcome is the endosomal escape. Obviously, if the complex escapes the endosome too early, it may never reach the underlying cell regions. The escape must

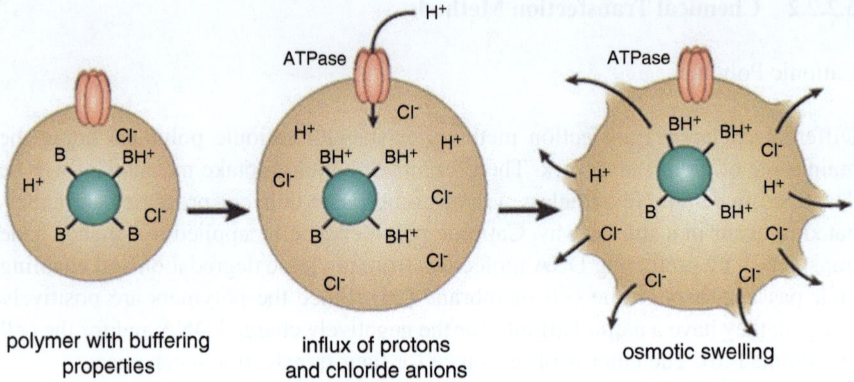

Fig. 6.5 Schematic representation of the proton sponge hypothesis: protons and chloride anions enter into the endosome, leading to osmotic swelling and finally to endosome rupture [29]

therefore be delayed until the nucleus is approached [27]. On the other hand, it is necessary that the system leaves the endosome prior to encountering the lysosomes, since these organelles contain enzymes that are able to degrade DNA. Many polyplexes exploit the low endosomal pH (5–5.5) to trigger endosomal membrane activity and permit their entry into the cytoplasm [6]. One well-defined principle is called the "proton sponge hypothesis" and applies to several cationic as well as anionic polymers [27].

A schematic representation of the abovementioned theory is shown in Fig. 6.5. The proton sponge hypothesis suggests that the polymers are protonated in the endosome buffer because of the reduced endosomal pH. The latter results in expansion of the polymer network by internal charge repulsion. In addition, the pH disturbance causes membrane located ATPase proton pumps to induce a proton influx into the endosome. This proton influx is accompanied with an influx of chloride counter anions causing osmotic swelling of the endosome. The simultaneous occurrence of these two phenomena ultimately leads to endosomal rupture. The latter phenomena reduce endosomal life expectancy significantly and therefore cause earlier release of the polyplexes into the cytoplasm [28]. Once the complexes are released, they must overcome additional barriers in the cytosol that hamper the complex delivery into the nucleus of the host cell.

Cytoskeletal elements that are present in the viscous cytoplasm function as molecular sieves and prevent the diffusion of large molecules. As a result, the diffusion of free DNA bases in the cytosol is negligible. Cationic polymers that compact DNA into small particles should aid in the movement of the DNA to the nucleus. It is anticipated that the complexes may exert some nuclear-localization effects and offer protection for DNA from fragmentation in the cytoplasm. Furthermore, polyplexes require dissociation prior to DNA delivery into the nucleus. Therefore, upon arriving within the nucleus, some enzymes associated with transcription are able to release the DNA from the cationic polymer [27].

The last barrier that has to be overcome is the nuclear membrane in order to gain access to the transcriptional machinery of the nucleus [7]. This membrane prevents the cellular uptake of most macromolecules larger than 70 kDa, unless they are able to interact with the nuclear pore active transport system. Pore complexes within the nuclear envelope enable trafficking of the DNA complexes between the cytoplasm and the nucleus, particularly when the DNA contains specific sequences recognized by host cell transcription factors [30]. Another possible route for DNA transport is that the DNA becomes physically associated with chromatin during mitosis when the nuclear envelope breaks down [12]. As soon as the therapeutic DNA is internalized in the nucleus, gene expression can be obtained via transcription and translation (see Sect. 6.1).

Overview of Most Important Cationic Polymers for DNA Delivery

An extended series of polymers with varying chemical structures and properties have already been developed, studied, and compared to be applied as DNA delivery systems. In the following paragraphs, some of the most important examples of such polymers are very briefly reviewed.

The main synthetic cationic polymer that is extensively used for gene delivery is poly(ethyleneimine) (PEI). It is synthesized in both linear and branched form and exists in different molecular weights [26]. Synthesis of linear PEI proceeds via ring opening polymerization of 2-ethyl-2-oxazoline followed by hydrolysis, whereas branched PEI is synthesized via acid-catalyzed polymerization of aziridine [7].

Another often studied polymer for polyplex formation is poly(L-lysine) (PLL). It is one of the first polymers of which it was demonstrated that it exhibits the capability to condense DNA. Due to neighboring group effects in the polymer chains, not all of the primary ε-amino groups of PLL are protonated at physiological pH [7]. Therefore, PLL shows a rather low buffering capacity (pH 5.7–7.7) and consequently only a low transfection efficiency [31]. In addition, it has been shown that PLL is cytotoxic and protects DNA from degradation only to a minor extent. Accordingly, PLL is commonly applied as a control in studies to demonstrate the advantages and disadvantages of other polyplexes [26].

One of the most important pH responsive polymers studied up to now is poly(2-(N,N-dimethylamino)ethyl methacrylate) (pDMAEMA). Due to its inherent cationic charge, it offers potential as gene transfer agent [6]. Synthesis of pDMAEMA occurs via free radical polymerization of 2-(N,N-dimethylamino)ethyl methacrylate (DMAEMA) (see Sect. 3.1 of Chap. 2). Evaluation of pDMAEMA indicated a combination of appreciable transfection efficiency and acceptable cytotoxicity [7]. Since the polymer destabilizes endosomes and dissociates easily from the DNA once delivered into the cytosol, pDMAEMA polyplexes have shown to be successful with respect to in vitro transfection efficiency [32]. Moreover, various modifications of the polymer structure have already been investigated in an attempt to further improve its transfection efficiency.

Linear poly(amidoamine) (PAA) polymers are known to have good DNA binding characteristics and are efficient gene transfer agents. Both cationic and amphoteric PAA derivatives exist and both of them have already been investigated for gene transfer applications [7]. The in vitro cytotoxicity of amphoteric PAA structures is low in comparison with PEI and PLL, due to the presence of negative charges (carboxylate anions) in the polymer at physiological pH. For in vitro gene delivery, amphoteric PAA is of greater interest than its cationic counterpart [33].

Chitosan is a carbohydrate-based polymer of which its biodegradability, its biocompatibility, and its cationic appearance render it one of the most prominent, naturally derived non-viral vectors for gene transfer. Due to chain entanglement effects, high molecular weight chitosan forms stable complexes with DNA. In addition, chitosan has been demonstrated to be highly safe, showing long-term retention systemically. Therefore, the polymer is suitable for passive targeting to disease regions showing a highly vascular permeation [34].

Another commonly applied chemical transfection method is the use of dendrimer-based vectors. Several highly branched organic polymers are interesting compounds that bind DNA and deliver it to host cells. Poly(amidoamine) (PAA) dendrimers are the most commonly used dendrimer-based vectors for gene transfer due to their ease of production and their commercial availability. Various modifications of the basic PAA dendrimers, with respect to toxicity, complex formation, cell binding, endosomal release, and cell targeting, have already been investigated in an effort to improve transfection efficiency [7]. Other relevant dendrimer vectors are based on poly(propyleneimine) and poly(L-lysine). Finally, some carbosilane and phosphor-containing dendrimers are also of interest as possible delivery devices.

Figure 6.6 illustrates the chemical structures of the above-described polymers. There exist also other interesting carbohydrate-based polymers, including

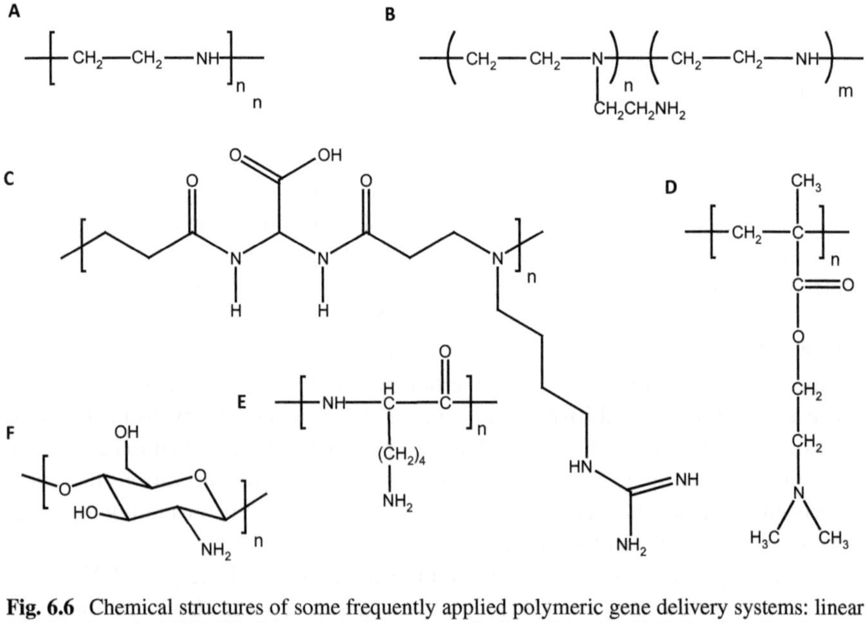

Fig. 6.6 Chemical structures of some frequently applied polymeric gene delivery systems: linear PEI (**a**), branched PEI (**b**), linear cationic amphoteric PAA (**c**), pDMAEMA (**d**), PLL (**e**), and chitosan (**f**)

β-cyclodextrin, schizophyllan, and modified dextran. In addition, various other degradable polymers have already been synthesized and evaluated for gene delivery purposes. Some of the most common examples include polyamino acids, polyaminoesters, and phosphor-containing polymers [7].

Alternative Chemical Transfection Methods

In addition to cationic polymers, several other chemical transfection methods also exist. Some of the most important vector systems are briefly described in the following paragraphs.

The working principle of lipid-based vectors is similar to that of polymer-based vectors. Liposome-mediated gene transfer was one of the earliest strategies applied to introduce exogenous genetic material into host cells. It has been shown that liposomes have a significant fusogenic potential with cell membranes and that they can form relatively stable complexes with DNA [7]. These complexes are often referred to as lipoplexes, and the gene transfection process is commonly called lipofection. The liposomes used for lipoplex formation typically contain at least two components: an amphiphilic cationic lipid and a neutral helper lipid. Cholesterol and DOPE are often used as neutral lipids. When the cationic lipid is laterally phase-separated by interaction with negatively charged macromolecules, the neutral lipid may form non-bilayer structures [35]. This may facilitate the destabilization of the cellular membranes and thereby mediate the cytoplasmic delivery of the therapeutic DNA.

The use of calcium phosphates is one of the cheapest and easiest methods to chemically deliver genes to cells. The process is based on the usage of divalent cations for bacterial transformation [36]. In this method, the therapeutic DNA is first diluted in a calcium chloride solution. This solution is then combined with a phosphate-containing buffered saline solution. As a result, a fine precipitate consisting of calcium cations and phosphate anions is formed, binding the therapeutic DNA on its surface. Finally, the precipitate suspension is delivered to the host cells, which take up some of the DNA-containing precipitate via endocytosis [37]. The precipitate protects the DNA from serum nucleases and ensures a stable and sustained expression of the therapeutic gene. An important disadvantage of the use of calcium phosphates for gene therapy applications is that the precipitate particles cover a large range of sizes, whereby synthesis is not reproducible [38].

Oligopeptides can also be used as agents for chemical transfection. Peptide-DNA conjugates offer a unique strategy for gene delivery into cells with high efficiency and cell specificity. These peptide vectors are able to deliver oligonucleotides into cells by utilizing short sequences of basic amino acid residues. These residues can readily cross the plasma membrane and thereby provide an efficient outset of the transfection process [39]. Examples of interesting oligopeptide vectors include the antennapedia homeodomain, the N-methylpurine-DNA glycosylase (MPG), and the transportan peptides [7].

In order to complete the overview of the vector systems used for chemical transfection, it has to be mentioned that nanoparticles can also be used as gene transfer vectors. Examples of some applicable nanoparticles are quantum dots, gold nanoparticles (see gene gun, Sect. 6.2.2.1), silica nanoparticles, carbon nanotubes, lipid-based nanoparticles, and polymeric hydrogels [7].

During the past decade, many different gene delivery systems have already been applied, from which plenty of structure-property relations have emerged. Furthermore, tremendous progress has been made in the design and synthesis of non-viral vectors for gene delivery. For some applications, very suitable vector systems have already been developed. Nevertheless, since vector characteristics are advantageous for some purposes but disadvantageous for others, up to now no universal vector serving all purposes has been discovered.

6.3 Scaffold-Mediated Gene Delivery

Recently, in an attempt to increase the efficiency of certain polymer-based delivery systems, polyplexes have been combined with polymeric scaffolds [26]. The gene delivery is then referred to as scaffold mediated. Figure 6.7 shows a graphical representation of the combined system of plasmid DNA-based polyplexes and polymeric scaffolds.

The strategy of scaffold-mediated gene delivery is advantageous for gene transfer due to several reasons. It enables localized delivery of the therapeutic gene and

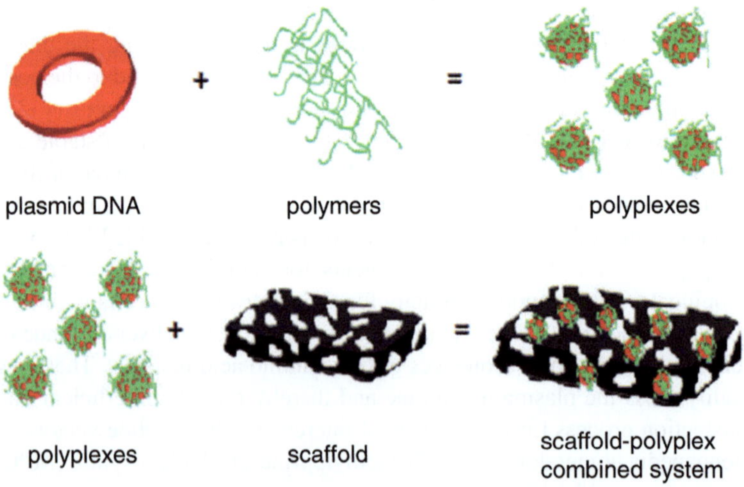

Fig. 6.7 Graphical representation of the combination of polyplexes and polymeric scaffolds for scaffold-mediated delivery in gene therapy [26]

thereby limits the unwanted exposure to other areas [40]. For biodegradable scaffolds, the genetic material is gradually released upon scaffold degradation. Interestingly, since the degradation rate of the scaffold material can be tailored, it is possible to regulate the release rate of the incorporated genes [26]. Other advantages of scaffold mediation include the protection of the polyplexes against clearing and enzyme degradation by the surrounding scaffold material and a reduced inflammatory response. Furthermore, scaffolds can act as bioactive agents for cellular recruitment or they can be preloaded with cells in situ [41]. Finally, scaffolds can serve as a platform for tissue regeneration, by providing a (load bearing) structure that creates and maintains a space, supports cell adhesion, and organizes tissue formation. This implies that scaffold-mediated gene delivery is a very interesting technique for plenty of tissue engineering applications [42].

In scaffold-mediated tissue engineering, therapeutic DNA, which is complexed with a certain vector, is delivered together with its scaffold at the site of the targeted tissue in order to provide conduction and induction [10]. Conduction refers to the purpose of the matrix to maintain a space and to provide a physical support for tissue regeneration [43]. Adversely, induction defines the role of the matrix as a drug delivery system to stimulate tissue formation. The introduction of therapeutic genes in cells can be performed in two different ways. In a first way (i.e., the in vivo approach), it occurs through transfection by plasmid DNA from the scaffold, whereas in a second manner (i.e., the ex vivo approach) gene introduction occurs via in vitro genetically manipulated cells prior to seeding onto the scaffold to be transplanted into the host tissue [10].

Numerous different scaffold geometries and sizes have already been developed for gene delivery applications [26]. Depending on the targeted clinical application, sizes from the macroscale up to the nano-level have been utilized. In addition, different morphologies have also been evaluated, including films, spheres, pellets, and sponges [44]. The degradation rate, the pore size, and the mechanical stiffness of the scaffolds can be tailored for specific applications by varying several (processing) parameters of the scaffolds.

In addition to the shape and the size of the scaffold, the applied material is also of major importance. Therefore, a considerable amount of polymeric materials has already been investigated for their potential use as gene delivery vehicles. A distinction can be made between hydrophilic materials including hyaluronic acid, collagen, and poly(ethylene glycol) and more hydrophobic materials such as poly(lactide-co-glycolic acid) [10]. Furthermore, scaffold materials can be categorized as natural, synthetic, or a combination thereof (natural-synthetic combinations) [45]. In the following sections, the abovementioned possibilities and some common examples will be concisely discussed. In addition, the principle of protein immobilization on the surface of thin films to be applied as scaffold material will be explained. Figure 6.8 shows a schematic representation of the principle of scaffold-mediated gene delivery.

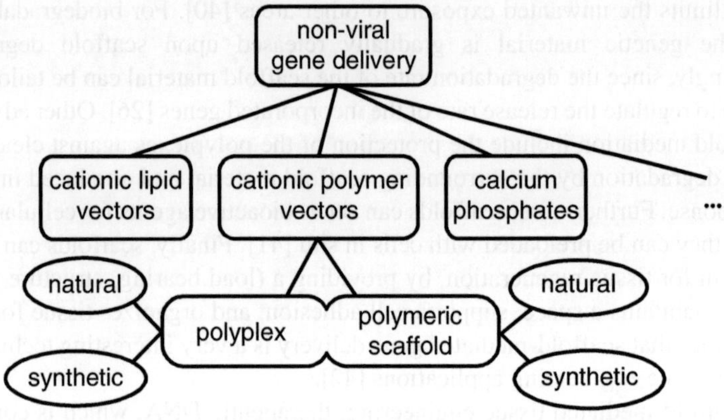

Fig. 6.8 Concept diagram of the combination of polyplexes and polymeric scaffolds for scaffold-mediated delivery in gene therapy

6.3.1 Natural Polymeric Scaffolds

Since the main principle in scaffold design is to mimic the natural environment, natural materials are an obvious choice for scaffold development. The main advantages of most natural scaffolds compared to synthetic scaffolds include their safety, their lower in vivo toxicity and immune response, and their cell-interactive properties [46]. A possible drawback is that natural scaffolds generally show poor mechanical properties. However, since in many cases the natural polymers can be cross-linked to improve these properties, this drawback may be readily eliminated. Finally, natural polymeric scaffolds normally exhibit a fast degradation rate. However, depending on the targeted application, the degradation rate can be decreased by modifying the polymers with higher amounts of cross-linkable moieties [26].

For tissue engineering purposes, collagen is the most widely applied natural material. Collagen has already been processed into the more manageable and less immunogenic form of atelocollagen, which can be processed into a variety of interesting derivatives [47]. Another natural polymer that is frequently utilized as scaffold material is agarose. This biopolymer is a linear polysaccharide which forms a thermoreversible gel when dissolved in water. The gels are non-immunogenic and noninflammatory, and they can easily be injected into specific disease sites [48]. Moreover, they exhibit highly effective polyplex release with large transfection rates. Possible drawbacks include their poor mechanical properties and possible difficulties in material handling. An excellent review from Yin et al. covering all protein-based therapeutic delivery agents could be consulted for a complete overview of protein-mediated gene delivery [49].

The combination of two alternative natural polymers (i.e., chitosan and alginate) results in a strong polyelectrolyte complex and enables the formation of a stable scaffold. Chitosan has the benefit that it possesses a low cytotoxicity and that it is nonantigenic. On the other hand, alginate can form gels easily and shows superior mechanical properties over chitosan [50]. Other examples of natural polymer materials to be applied as scaffolds for gene delivery include gelatin, fibrin, and glycosaminoglycans [45]. These polymers and derivatives thereof are extensively examined in order to enhance gene transfer processes. Clinical investigations are ongoing and preliminary results seem very promising [51].

6.3.2 Synthetic Polymeric Scaffolds

In addition to natural polymeric scaffolds, also synthetic polymeric scaffolds are suitable to be applied as gene delivery scaffolds. A very important benefit of synthetic scaffolds is the ease of creating a material with both tailor-made mechanical as well as chemical properties. The latter implies that degradation properties, mechanical stiffness and strength, porosity, and surface functional properties can be precisely designed for every scaffold application [10]. Interestingly, synthetic polymeric scaffolds can be produced in bulk with a high batch reproducibility, which is a very important consideration when designing an implant to be fabricated on a large scale [26].

The most extensively studied synthetic material for gene delivery applications is poly(lactic acid-co-glycolic acid) (PLGA). PLGA is a biodegradable polymer with a large application platform in the clinical practice [52]. A large variety of PLGA implants with multiple functionalities has already been developed. Up to now, PLGA films, disks, sponges, microspheres, and nanoparticles have been tested as polyplex reservoirs. Literature data indicate that the copolymer is non-cytotoxic and that it efficiently protects the DNA incorporated in the polyplexes from enzymatic degradation in vitro [26].

Another biodegradable synthetic polyester applied for scaffold-mediated delivery is poly(D,L-lactic acid) (PDLLA), which possesses similar properties as PLGA. An important disadvantage of PDLLA and PLGA is their acidic degradation products, which often lead to inflammatory responses in vivo [53]. Depending on the application and the desired degradation rate, porous PDLLA disks and films can be applied as scaffolds for DNA delivery. In addition to the above-described polyesters, a series of functionalized PDLLA has already been investigated for scaffold-mediated transfection applications [45].

Poly(ethylene glycol) (PEG) can also be used as hydrogels to encapsulate polyplexes and cells. PEG gels are highly flexible, show a very low cytotoxicity in vivo, and are noninflammatory [26]. These properties render PEG a very desirable synthetic polymer for gene delivery scaffolds and tissue engineering applications. However, regular PEG hydrogels are often too weak and therefore require crosslinking prior to their use in specific applications.

In addition to the abovementioned synthetic polymers, other compounds have also been studied, including poly(ethylene-co-vinylacetate), poly(orthoester)s, poly(β-amino ester), polyanhydrides, and polyurethanes [45].

6.3.3 Combinations of Natural and Synthetic Polymers

A final strategy for substrate-mediated gene delivery is the application of a blend of natural and synthetic polymers. The aim is to combine the benefits of natural polymers with those of synthetic polymers in one scaffold material. Although this approach is very interesting to date, only minor research has been performed on this topic.

One example of a scaffold that is composed of a combination of a natural and a synthetic polymer is a collagen/poly(glycolic acid) (collagen/PGA) scaffold. By the addition of PGA fibers to a solution of type I collagen and subsequent freeze-drying and cross-linking, a collagen/PGA sponge was prepared [26].

6.4 Applications

Gene therapy has emerged as a versatile and powerful tool to treat a wide variety of diseases. The initial concept of gene therapy was focusing on the treatment of genetic diseases, but during the last decades the focus has broadened considerably. More recent applications of gene therapy include tissue engineering, cancer therapy, and ocular gene therapy.

For gene therapy treatments to be successful, several requirements must be fulfilled. The most important one is that a potential gene of interest must be identified and cloned in an appropriate expression vector. Furthermore, after cell transfection, sufficient and controlled gene expression must be achieved. Other important issues include a clear understanding of the pathogenesis of the targeted disease and the possible side effects [54].

6.4.1 Genetic Diseases

A wide range of genetic diseases has already been treated using gene therapy. Genetic diseases can be monogenic but the majority is multifactorial and therefore rather difficult to treat using gene therapy. Some of the most common monogenic diseases include mucopolysaccharidosis, familial hypercholesterolemia, hemophilia B, Duchenne's muscular dystrophy, Gaucher's disease, and cystic fibrosis [54]. Since these diseases are caused by a mutation in one single gene, they are interesting candidates for gene therapy.

Cystic fibrosis (CF) is the most common lethal genetic disease in the Caucasian population. This lung disease is caused by a mutation in a gene that encodes for the cystic fibrosis transmembrane conductance regulator (CFTR) protein, which is expressed in many epithelial and blood cells. This protein possesses many regulatory roles, but it functions mainly as a chloride channel situated on the apical surface

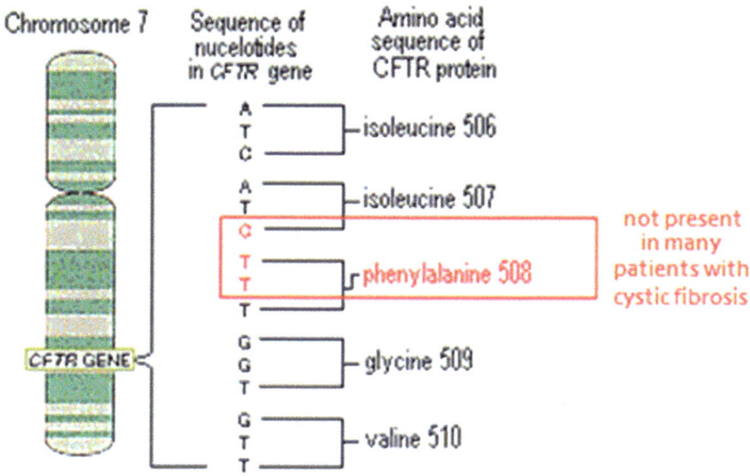

Fig. 6.9 Schematic representation of the CFTR gene and the deletion of phenylalanine at position 508 on chromosome 7, leading to CF [56]

of epithelial cells. Many CFTR gene mutations have already been identified, but the deletion of phenylalanine at position 508 on the long arm of chromosome 7 (see Fig. 6.9) accounts for about 65% of the mutated alleles in Northern European and North American populations [55]. A correction of this defect by the delivery of a normal CFTR gene to the lung could therefore prolong the life span of CF patients. In addition to the fact that CF is a monogenic disease, the target epithelial cells are directly accessible by noninvasive techniques which renders the disease suitable to be treated using gene therapy. Interestingly, the receptors for some viral vectors are in general situated on the inaccessible basolateral membrane and not on the apical membranes (where gene delivery is targeted) [54].

6.4.2 Tissue Engineering

Tissue engineering is defined as the science of generating new living tissues to replace, repair, or augment the tissue that is diseased or damaged. Tissue engineering has the aim to restore tissue or organ functions [57]. Three important requirements are associated with this application. A scaffold or matrix to support cells for tissue formation is needed. Next, responsive cells have to be present to produce compounds forming (extra-)cellular matrix. Finally, signaling factors promoting cell division, maturation, and differentiation of the cells are also required [2].

Gene therapy possesses different applications related to tissue engineering [58, 59]. An important example is wound healing of the skin [60]. Delayed wound healing can be caused by diseases such as diabetes and peripheral vascular disease or by steroid therapy. Wound repair is a complex cascade involving multiple interactions between cytokines, cells, and the extracellular matrix. The skin is an ideal candidate for gene therapy because of several reasons, including its accessibility and the short-term regulation of a multitude of cytokines and growth factors crucial for regeneration purposes [61]. Short-term gene expression is often a problem with gene therapy vectors, but it is desirable when transduction or transfection of the skin for wound repair is intended.

Other important gene therapy applications in the field of tissue engineering [62] include the regeneration of cartilage [63], bone [64, 65], nerves [66], liver, endocrine pancreas, muscle [67, 68], and blood vessels [10].

6.4.3 Cancer Therapy

Although major advances have been made in the field of cancer treatment, the disease is still one of the most important causes of death. Cancer tumors are formed when tissue homeostasis, regulated by different genes, is disturbed. Two different types of regulating genes exist, namely, growth-promoting oncogenes and growth-suppressing tumor suppressor genes [69]. Mutations resulting in either the excessive activation of oncogenes or the inactivation of growth-suppressing genes can ultimately lead to tumor development.

A lot of clinical research has already been performed on possible cancer treatments. In addition to current treatments including chemotherapy, different gene therapy approaches have also been tested in the laboratory [70]. A few of these approaches are at present being evaluated in the clinic. Some examples include suicide gene therapy, cytokine gene therapy (i.e., immunotherapy), and anti-angiogenesis therapy [54]. A last gene therapy approach applied in the clinical practice has the aim to either suppress oncogenes or restore the normal function of tumor suppressor genes [71].

6.4.4 Ocular Gene Therapy

Ocular gene transfer has gained increasing interest because of safety issues and efficacy characteristics. Visual function in humans can be hampered by different inherited as well as acquired diseases. An important example is age-related macular degeneration (ARMD), which is the most common cause of irreversible visual impairment for the elderly population [72]. The disease affects the macula, which is the part of the eye enabling the perception of fine details. As illustrated in Fig. 6.10, the macula is located in the center of the retina (i.e., the light-sensitive tissue at the back of the eye) [73]. The therapeutic effect of the current treatments for ARMD, including photocoagulation therapy and the administration of antioxidant supplements, is insufficient. Therefore, the high prevalence of the disease, together with

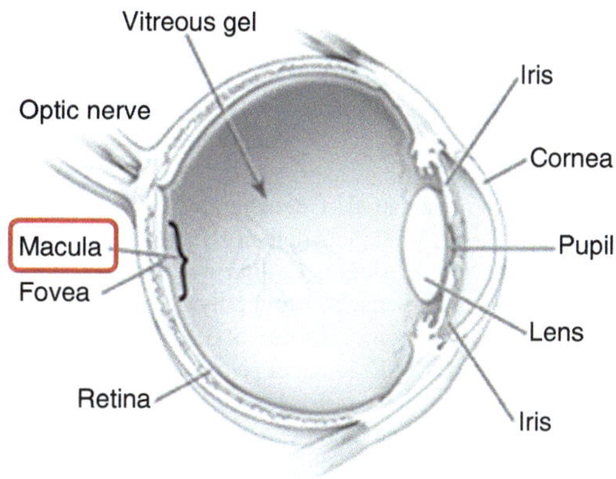

Fig. 6.10 Schematic representation of the human eye (cross-section) [73]

the fact that the age of the population is increasing, renders novel treatment approaches including gene therapy extremely interesting.

The eye is an organ particularly suitable for local gene therapy [74]. The latter is related to accessibility. In addition, because of its small size and enclosed structure, the required therapeutic drug concentration is relatively low, and the diffusion of active products from the eye into the circulation is minimal [75]. Moreover, a series of genes causing eye diseases have already been identified, which is an important requirement to enable gene therapy treatments. Finally, both rodents as well as large animal models resembling human eye pathologies are available [73].

Viral vectors including retroviruses, adenoviruses, and adeno-associated viruses have already been evaluated for ocular gene therapy applications. These vectors have elicited high levels of therapeutic protein expression in ocular cells and tissues [75]. However, since most blinding ocular diseases are not life-threatening and viral vectors can impose safety issues, the use of potentially harmful viral vectors remains questionable [76].

Non-viral vectors show significant advantages over viral carriers. In general, they are safe and nontoxic to ocular tissues. Although non-viral gene transfer efficiency has already been improved, the final transfection rate is still rather low, and the expression of the transgene shows a relatively short life span [73]. These drawbacks can be overcome by the use of slow release devices and, if needed, by repeated treatments.

Various methods to deliver genes to ocular cells have already been investigated. Several carriers have been used to enhance DNA stability, to ease the entry into the cells, and/or to allow slow gene release. Some examples include lipids, peptides, and various polymers such as poly(L-lysine), cyanoacrylates, poly(ethyleneimine), and gelatin. Furthermore, several types of nano- and microparticles and combinations thereof have been examined [75].

References

1. W.F. Anderson, Human gene therapy. Nature **392**, 25–30 (1998)
2. X.D. Sun, I.S. Lee, Gene technology in tissue engineering. Am. J. Biochem. Biotechnol. **2**(2), 66–72 (2006)
3. D.H. Abramson et al., Gene therapy. Genet. Home Ref. **1**(7), 123–131 (2010)
4. D.S. Strayer, Viral gene delivery. Expert Opin. Investig. Drugs **8**(12), 2159–2172 (1999)
5. E.N. Marieb, K. Hoehn, Cell growth and reproduction. Anat. Physiol. **3**, 92–105 (2008)
6. P. Dubruel, E. Schacht, Vinyl polymers as non-viral gene delivery carriers: current status and prospects. Macromol. Biosci. **6**(10), 789–810 (2006)
7. M.A. Mintzer, E.E. Simanek, Nonviral vectors for gene delivery. Chem. Rev. **109**(2), 259–302 (2009)
8. R. Miesfeld, Gene therapy and stem cell technology. J. Appl. Mol. Genet. **25**, 258–262 (2001)
9. M. Schleef et al., Production of non viral DNA vectors. Curr. Gene Ther. **10**(6), 487–507 (2010)
10. O. Bleiziffer et al., Gene transfer strategies in tissue engineering. J. Cell. Mol. Med. **11**(2), 206–223 (2007)
11. D.W. Hutmacher, A.J. Garcia, Scaffold based bone engineering by using genetically modified cells. Gene **347**(1), 1–10 (2005)
12. C.M. Wiethoff, C.R. Middaugh, Barriers to nonviral gene delivery. J. Pharm. Sci. **92**(2), 203–217 (2003)
13. A. Herschhorn, A. Hizi, Retroviral reverse transcriptases. Cell. Mol. Life Sci. **67**(16), 2717–2747 (2010)
14. W.S. Hu, V.K. Pathak, Design of retroviral vectors and helper cells for gene therapy. Pharmacol. Rev. **52**(4), 493–511 (2000)
15. A. Sharma et al., Adenovirus receptors and their implications in gene delivery. Virus Res. **143**(2), 184–194 (2009)
16. R.F. Doll et al., Comparison of promoter strengths on gene delivery into mammalian brain cells using AAV vectors. Gene Ther. **3**, 437–447 (1996)
17. X.J. Zhang, W.T. Godbey, Viral vectors for gene delivery in tissue engineering. Adv. Drug Deliv. Rev. **58**(4), 515–534 (2006)
18. R.H. Lachmann, S. Efstathiou, Use of herpes simplex virus type I for transgene expression within the central nervous system. Clin. Sci. **96**, 533–541 (1999)
19. J.C. Glorioso, N.A. Deluca, D.J. Fink, Development and application of herpes simplex virus vectors for human gene therapy. Annu. Rev. Microbiol. **49**, 675–710 (1995)
20. Y. Lei et al., Incorporation of active DNA/cationic polymer polyplexes into hydrogel scaffolds. Biomaterials **31**(34), 9106–9116 (2010)
21. J.A. Wolff et al., Direct gene transfer into mouse muscle in vivo. Science **247**, 1465–1468 (1990)
22. E. Neumann, S. Kakorin, K. Toensing, Fundamentals of electroporative delivery of drugs and genes. Bioelectrochem. Bioenerg. **48**(1), 3–16 (1999)
23. J. Hao et al., Gene delivery to cornea. Brain Res. Bull. **81**(2-3), 256–261 (2010)
24. B. Gonzalez et al., A novel synthetic strategy for covalently bonding dendrimers to ordered mesoporous silica: potential drug delivery applications. J. Mater. Chem. **19**, 9012–9024 (2009)
25. S.C. De Smedt, J. Demeester, W.E. Hennink, Cationic polymer based gene delivery systems. Pharm. Res. **17**(2), 113–126 (2000)
26. S. O'Rorke, M. Keeney, A. Pandit, Non-viral polyplexes: scaffold mediated delivery for gene therapy. Prog. Polym. Sci. **35**(4), 441–458 (2010)
27. C.W. Pouton, L.W. Seymour, Key issues in non-viral gene delivery. Adv. Drug Deliv. Rev. **46**(1-3), 187–203 (2001)
28. J.P. Behr, The proton sponge: a trick to enter cells the viruses did not exploit. CHIMIA Int. J. Chem. **51**(2), 34–37 (1997)

29. D.W. Pack et al., Design and development of polymers for gene delivery. Nat. Rev. Drug Discov. **4**, 581–593 (2005)
30. D.A. Dean, Import of plasmid DNA into the nucleus is sequence specific. Exp. Cell Res. **230**(2), 293–302 (1997)
31. P. Dubruel, L. Dekie, E. Schacht, Poly-L-glutamic acid derivatives as multifunctional vectors for gene deliver. Part A. Synthesis and physicochemical evaluation. Biomacromolecules **4**, 1168–1176 (2003)
32. P. van de Wetering et al., Structure-activity relationships of water-soluble cationic methacrylate/methacrylamide polymers for nonviral gene delivery. Bioconjug. Chem. **10**(4), 589–597 (1999)
33. P. Ferruti et al., Amphoteric linear poly(amido-amine)s as endosomolytic polymers: correlation between physicochemical and biological properties. Macromolecules **33**(21), 7793–7800 (2000)
34. Y. Kato, H. Onishi, Y. Machida, Biological characteristics of lactosaminated N-succinyl-chitosan as a liver-specific drug carrier in mice. J. Control. Release **70**(3), 295–307 (2001)
35. C.T. de Ilarduya, Y. Sun, N. Duezguenes, Gene delivery by lipoplexes and polyplexes. Eur. J. Pharm. Sci. **40**(3), 159–170 (2010)
36. F.L. Graham, A.J. Vandereb, A new technique for the assay of infectivity of human adenovirus 5 DNA. Virology **52**(2), 456–467 (1973)
37. A. Kovtun, R. Heumann, M. Epple, Calcium phosphate nanoparticles for the transfection of cells. Biomed. Mater. Eng. **19**(2-3), 241–247 (2009)
38. P.L. Felgner, Advanced drug delivery. Adv. Drug Deliv. Rev. **5**, 163–187 (1990)
39. J.H. Wang, X.Y. Liu, Targeting strategies in cancer gene therapy. Acta Biochim. Biophys. Sin. **35**(4), 311–316 (2003)
40. J.H. Jang, C.B. Rives, L.D. Shea, Plasmid delivery in vivo from porous tissue-engineering scaffolds: transgene expression and cellular transfection. Mol. Ther. **12**(3), 475–483 (2005)
41. J.H. Jang, T.L. Houchin, L.D. Shea, Gene delivery from polymer scaffolds for tissue engineering. Expert Rev. Med. Devices **1**(1), 127–138 (2004)
42. J. Bonadio, Tissue engineering via local gene delivery. J. Mol. Med. **78**(6), 303–311 (2000)
43. J.H. Jang et al., Surface adsorption of DNA to tissue engineering scaffolds for efficient gene delivery. J. Biomed. Mater. Res. A **77A**(1), 50–58 (2006)
44. Y.G. Lei, T. Segura, DNA delivery from matrix metalloproteinase degradable poly(ethylene glycol) hydrogels to mouse cloned mesenchymal stem cells. Biomaterials **30**(2), 254–265 (2009)
45. H.L. Fu, S.X. Cheng, R.X. Zhuo, Biodegradable polymers for controlled release of gene delivery systems. Acta Polym. Sin. **2**, 97–103 (2009)
46. B.G. Sengers et al., Computational modelling of cell spreading and tissue regeneration in porous scaffolds. Biomaterials **28**(10), 1926–1940 (2007)
47. T. Ochiya et al., Atelocollagen-mediated drug discovery technology. Expert Opin. Drug Discov. **2**(2), 159–167 (2007)
48. J.M. Nancy et al., Sustained release of plasmid DNA using lipid microtubules and agarose hydrogel. J. Control. Release **88**(2), 321–331 (2003)
49. L.M. Yin, C. Yuvienco, J.K. Montclare, Protein based therapeutic delivery agents: contemporary developments and challenges. Biomaterials **134**, 91–116 (2017)
50. P.B. Malafaya, G.A. Silva, R.L. Reis, Natural-origin polymers as carriers and scaffolds for biomolecules and cell delivery in tissue engineering applications. Adv. Drug Deliv. Rev. **59**(4-5), 207–233 (2007)
51. Z.H. Wang et al., Cartilage tissue engineering with demineralized bone matrix gelatin and fibrin glue hybrid scaffold: an in vitro study. Artif. Organs **34**(2), 161–166 (2010)
52. O.V. Chumakova et al., Composition of PLGA and PEI/DNA nanoparticles improves ultrasound-mediated gene delivery in solid tumors in vivo. Cancer Lett. **261**(2), 215–225 (2008)

53. M. Heyde et al., Development of a slow non-viral DNA release system from P(DL)LA scaffolds fabricated using a supercritical CO2 technique. Biotechnol. Bioeng. **98**, 679–693 (2007)
54. A. Mhashilkar et al., Gene therapy—therapeutic approaches and implications. Biotechnol. Adv. **19**(4), 279–297 (2001)
55. B.P. O'Sullivan, S.D. Freedman, Cystic fibrosis. Lancet **373**, 1891–1904 (2009)
56. M.J. Welsh, A.E. Smith, Cystic fibrosis. Sci. Am. J. **2**, 1–6 (1995)
57. A. Curtis, M. Riehle, Tissue engineering: the biophysical background. Phys. Med. Biol. **46**(4), 47–65 (2001)
58. J.S. Park et al., Multimodal magnetic nanoclusters for gene delivery, directed migration, and tracking of stem cells. Adv. Funct. Mater. **27**(25), 8 (2017)
59. T. Perisic et al., Biodegradable poly (lactic acid-co-glycolic acid) scaffolds as carriers for genetically modified fibroblasts. PLoS One **12**(4), e0174860 (2017)
60. P.S. Rabbani et al., Novel lipoproteoplex delivers Keap1 siRNA based gene therapy to accelerate diabetic wound healing. Biomaterials **132**, 1–15 (2017)
61. A.J. Singer, R.A. Clark, Cutaneous wound healing. English J. Med. **341**, 738–746 (1999)
62. K. Maeda et al., Optimizing the host substrate environment for cardiac angiogenesis, arteriogenesis, and myogenesis. Expert Opin. Biol. Ther. **17**(4), 435–447 (2017)
63. I.N. Aguilar et al., Customized biomaterials to augment chondrocyte gene therapy. Acta Biomater. **53**, 260–267 (2017)
64. S. Pacelli et al., Controlling adult stem cell behavior using nanodiamond-reinforced hydrogel: implication in bone regeneration therapy. Sci. Rep. **7**, 15 (2017)
65. M. Bez et al., In situ bone tissue engineering via ultrasound-mediated gene delivery to endogenous progenitor cells in mini-pigs. Sci. Transl. Med. **9**(390), eaal3128 (2017)
66. J.J. Siu et al., Improved gene delivery to adult mouse spinal cord through the use of engineered hybrid adeno-associated viral serotypes. Gene Ther. **24**(6), 361–369 (2017)
67. W.M. Han, Y.C. Jang, A.J. Garcia, Engineered matrices for skeletal muscle satellite cell engraftment and function. Matrix Biol. **60-61**, 96–109 (2017)
68. M. Rabbani et al., Cyclic stretch effects on adipose-derived stem cell stiffness, morphology and smooth muscle cell gene expression. Tissue Eng. Regen. Med. **14**(3), 279–286 (2017)
69. C. Theillet, Towards an inventory of oncogenic mutations in cancer. Bull. Cancer **97**(11), 1223–1229 (2010)
70. D. Haddad, Genetically engineered vaccinia viruses as agents for cancer treatment, imaging, and transgene delivery. Front. Oncol. **7**, 96 (2017)
71. B. Sangro, J. Prieto, Gene therapy for liver cancer: clinical experience and future prospects. Curr. Opin. Mol. Ther. **12**(5), 561–569 (2010)
72. B. Shastry, Genetic diversity and medicinal drug response in eye care. Graefes Arch. Clin. Exp. Ophthalmol. **248**(8), 1057–1061 (2010)
73. P. Colella, G. Cotugno, A. Auricchio, Ocular gene therapy: current progress and future prospects. Trends Mol. Med. **15**(1), 23–31 (2009)
74. J.W.B. Bainbridge, M.H. Tan, R.R. Ali, Gene therapy progress and prospects: the eye. Gene Ther. **13**(16), 1191–1197 (2006)
75. C. Andrieu-Soler et al., Ocular gene therapy: a review of nonviral strategies. Mol. Vis. **12**(150), 1334–1347 (2006)
76. N. Provost et al., Biodistribution of rAAV vectors following intraocular administration: evidence for the presence and persistence of vector DNA in the optic nerve and in the brain. Mol. Ther. **11**, 275–283 (2005)

Index

J. Van Hoorick et al. (eds.), *Polymer and Photonic Materials Towards Biomedical Breakthroughs*, Micro- and Opto-Electronic Materials, Structures, and Systems, https://doi.org/10.1007/978-3-319-75801-5

R

MIX
Papier aus verantwortungsvollen Quellen
Paper from responsible sources
FSC® C105338

If you have any concerns about our products,
you can contact us on
ProductSafety@springernature.com

In case Publisher is established outside the EU,
the EU authorized representative is:
Springer Nature Customer Service Center GmbH
Europaplatz 3, 69115 Heidelberg, Germany

Printed by Libri Plureos GmbH
in Hamburg, Germany